The Smart and Resilient City

The Smart and Resilient City

Models and Ways of Life

Edited by
Gilles N'Goala
Florence Paulhiac Scherrer
Fabien Durif

ISTE

WILEY

First published 2025 in Great Britain and the United States by ISTE Ltd and John Wiley & Sons, Inc.

ISTE Ltd
27-37 St George's Road
London SW19 4EU
UK

www.iste.co.uk

John Wiley & Sons, Inc.
111 River Street
Hoboken, NJ 07030
USA

www.wiley.com

Library of Congress Control Number: 2025935380

British Library Cataloguing-in-Publication Data
A CIP record for this book is available from the British Library
ISBN 978-1-78630-998-3

The manufacturer's authorized representative according to the EU General Product Safety Regulation is Wiley-VCH GmbH, Boschstr. 12, 69469 Weinheim, Germany, e-mail: Product_Safety@wiley.com.

Contents

Preface

This collective publication is the result of the work of the International Chair in Smart City Usages and Practices (CIT.us), a chair that since 2019 has brought together researchers from the School of Management Sciences of the University of Quebec at Montreal (ESG UQAM) and Montpellier Management from the University of Montpellier[1]. This chair has three objectives: (i) to produce research in the humanities and social sciences (public planning, urbanism, marketing, sociology, etc.), which gives the smart city a more human texture; (ii) to increase exchanges between the academic world and the ecosystem of smart cities (local authorities, large, medium and small businesses, networks of associations, etc.); and (iii) disseminate knowledge to communities of students and stakeholders, as well as to the general public, through courses, transfer tools (video clips, podcasts, barometers, etc.) and events (conferences, roundtables, etc.).

This collective publication is the result of study and research work conducted in the field, "in the real world", in collaboration with elected officials and technicians from the cities of Montpellier and Montreal in charge of the transformation of the city and the smart city. We warmly thank all of our partners, public and private, for the numerous exchanges and knowledge they have been able to share with us. There are too many for us to name them all.

This collective publication has been principally written by a diverse academic community, but is not solely theoretical and academic. The professional expertise of Jacques Priol, Philippe Sajhau, Anne Le Cabec and Alain Goudey informs this text, putting their practices into perspective. Their contribution thus helps us to develop a

Preface written by Gilles N'GOALA, Florence PAULHIAC SCHERRER and Fabien DURIF.

1 See: https://villeintelligente.esg.uqam.ca and https://chairecitus.com.

practical analysis of the smart city and highlights the challenges they face on a daily basis.

This collective publication aims to guide public and private decision-making by contributing to supporting critical reflection and improving the management of smart city projects.

– First, it adopts the human point of view and counterbalances the technicist approaches that often dominate in the world of smart cities. In terms of innovation, the human is often neglected in favor of technologies (sensors, artificial intelligence tools, etc.), in R&D investment, for example, while the transformation of the city, practices and uses is impossible without a good understanding of and effective support for the citizens and users who make up the territory (social acceptability, perceived transparency, etc.).

– Second, it gives a cross-sectional vision of the smart city, whereas organization by profession/area of activity (waste, mobility, food supply, etc.) frequently leads to seeing the city "in silos" and neglecting potential interactions between sectors. For example, improving the food supply in cities requires integrating issues of logistics (transport), distribution (short supply chains), urbanism (food and agro-ecological landscapes) or education (students, for example).

– Finally, it does not see it itself as "giving lessons" to the political personnel and professionals whose daily work this is and who have the heavy responsibility of transforming cities and territories. Its main aim is to encourage greater reflexivity and a true critical thinking, in the aim of avoiding pitfalls and failures (such as the Quayside project in Toronto, for example), especially when projects go beyond the experimental framework and are "scaled up" in the city.

We wish you happy reading.

Gilles N'GOALA
Florence PAULHIAC SCHERRER
Fabien DURIF
May 2025

Introduction

Smart City: Lives in Transition

I.1. Introduction

The smart city is intimately linked to the development of digital infrastructures that aim, as in the case of the metropolitan area of Angers, to better manage water resources (based on atmospheric conditions), connectivity (access to public Wi-Fi), public lighting (illumination, extension, regulation), waste management (collection and preventing dumping), mobility (urban traffic and parking) or safety and tranquility in public spaces.

Many cities have particularly invested in the development of hypervisors for better centralization of all monitoring tools, applications, reference points and data. For example, since April 2019, the OnDijon control center has been collecting data from the whole set of urban services, as well as from the metropolitan area's 300 video surveillance cameras and the city's sensors[1]. The city of Marseille collects data on its wooded areas using satellite images and feedback from its agents. Since 2019, the Nantes metropolitan area has focused on overall improvements for the population including issues around pollution, noise, heat peaks, food supply, physical and mental health, culture and sport. It therefore developed the Synopse project (*Système numérique d'observation populationnelle santé environnement* (digital system for population health and environment observation)) to aggregate these data.

Introduction written by Gilles N'GOALA, Florence PAULHIAC SCHERRER and Fabien DURIF.

1 *Le Monde* article: Van Eeckhout, L. (2023). "Ce que la "smart city" apprend aux villes"; "Que savent les maires de leur ville 1/3 ?". On the occasion of the October 17 "Le Monde Cities" conference in Marseille, October 9, 2023.

Driven by digital giants (IBM, Cisco, Google, etc.), smart city digital devices have developed in silence and spread unevenly, mainly affecting metropolises and leaving out peri-urban and rural areas (Ait-Yahia et al. 2019). The aim was first to save on resources (water, lighting and energy) and to provide better services to make life easier for the local population: 24/7 access, geolocation, real-time information, public Wi-Fi, open access to information, etc. The service city has taken shape through numerous technological applications. For example, the Internet of Things (IoT) and other sensors make people's lives easier through connecting everyday objects with the Internet (personal assistants, for example). Algorithms and artificial intelligence (AI) enable greater personalization of services, content and experiences for citizens. Open Data, which refers to data that everyone can access, use or share, allows multiple actors (large and small businesses, researchers, etc.) to have a better understanding of practices and uses within public space. Facial recognition, based on very private data (facial features), authenticates people and could help to guarantee greater security and tranquility in cities. But the servicization of the city through ever more advanced technologies is no longer the only objective. The climate emergency and sustainability issues have redirected priorities toward environmental preservation (water management, zero waste, etc.) and the fight against global warming (renaturation of cities, net zero artificialization, low-emissions zones, etc.).

Two visions of the city, technology and society are sometimes diametrically opposed, with:

– on the one hand, that in favor of a "detechnologization" of the world, believing that a decrease in technology in our lives is the only solution to respond to the climate emergency and to safeguard democracy;

– on the other hand, that in favor of technologies that may be useful for dealing with major contemporary problems, whether they are economic, ecological, social or democratic, provided that they are designed with this in mind (smart green tech, tech for good, responsible AI, etc.) (N'Goala and Collin-Lachaud 2023).

The concept of the "smart city" is often vilified, debased, scorned or denounced. Techno-centric or techno-solutionist approaches have had their day and people who can, consciously or unconsciously, still defend them are rare. Technology is much less important than people's lives and cannot claim to make them happy. Many people prefer epithets such as the inclusive and sustainable city, the citizen and participatory city, the smart city with positive impact, the learning city (UNESCO Global Network of Learning Cities), the service city, the innovative and connected city, the resilient and adaptive city, the sustainable and responsible city, or even the open and collaborative city. The city thus conveys an imagination, a vision, representations and values which sometimes oppose and often complement each other. The smart city is a realizable utopia that takes shape in territories and requires the establishment of digital infrastructures. Public and private decision-makers then assign multiple objectives and responsibilities to it (sustainability, social inclusion, user services, etc.). We will therefore use the original term "smart city", as it serves as a point of attention and reference in professional and academic circles.

The history of innovation is punctuated by periods of acceleration that mark both major technical and scientific advances and profound transformations in lifestyles. The intrusion of technology, particularly digital technology, into public spaces is in fact causing ruptures in traditions, whose scope is still difficult to measure. Let us imagine for a moment a total blackout of digital infrastructure – similar to what was imagined in 2023 in the Netflix disaster movie, *Leave the World Behind* – our lives would be dramatically turned upside down. This dependence on technology justifies not only questioning and criticizing this headlong rush, but also considering the upheavals it is causing in our lives.

Going hand in hand with globalization, the growing urbanization and metropolization of the world appear as inevitable phenomena that are sure to produce many negative externalities (air and water pollution, city congestion, social inequalities, etc.). While in 2019, more than 4 billion people lived in cities (55% of the world population), this number is expected to grow to 6.5 billion by 2050. In such a context, how can we maintain the individual and collective well-being of the inhabitants? How can we avoid a deterioration in the quality of life of billions of people? The well-being of populations is based, from a hedonic perspective, on the obtaining of pleasures, positive affects and the achievement of personal happiness and satisfaction in life, or from a eudaimonic perspective, on self-realization, in accordance with their true nature and the deep goals they pursue. Following this second perspective, Deci and Ryan (1985) consider that people are self-realized when their fundamental needs for autonomy, competence and relationships with others (interpersonal affiliation) are met. These needs motivate and drive human activities and practices and give meaning to them.

Is the smart city synonymous with happiness and self-realization for urban populations? It certainly offers them what they want: services, convenience (real-time information, 24/7 accessibility, time saving, less effort, etc.), more leisure activities and pleasures. And yet, will their individual and collective well-being be greater? How and to what extent will the cities of the future meet the fundamental needs of autonomy, competence and relationships with others? Does the expansion of technologies not further endanger the autonomy of local populations, their competences and their relationships with others?

Technologies do not guarantee that everyone will be self-realized and flourish in the cities of the future. First, the digital world evokes fears of a loss of control by residents over their living environment and poses a risk of personal alienation (powerlessness) or surveillance and social control by states, public authorities or digital giants (Google, Apple, etc.). Unlike a world where everyone can decide on their activities and choices in complete autonomy, the smart city entails standards, rules and values, a morality that imposes itself and constrains behavior, even if it is supposed to be for "the common good". Second, digital technology confronts people with their incompetence, making massive use of digital channels (mobile applications, sites, connected objects) and advanced data exploitation technologies (AI), while a large part of the population is marked by digital illiteracy (digital exclusion) and vulnerability in matters of privacy (intrusion, dissemination and theft of personal data). Finally, robotization and automation may lead to a reduction in social contact in favor of contact mediated by machines (bots, etc.) and increase social isolation among the population. For economic reasons, humans are increasingly being replaced by machines in managing social and commercial interactions. Things as mundane as discussions with the postman, the store clerk or the municipal employee are becoming rarer in everyday life. Relationships with others could thus diminish and loneliness could increase further, as well as the feeling of unease.

This book proposes to explore the constitutive dimensions of the so-called smart city, from one angle in particular – that of uses – and more broadly of the practices in which these innovations engage and the needs that they satisfy. City life is multi-faceted and the smart city intervenes in the daily lives of urban populations, from their homes to public spaces, but also in the institutions they frequent, whether cultural or health-related, for example, or in the media or transport. What are the impacts of these innovations on the quality of urban life, and how to account for the societal transformations they generate? These questions guide this work, inviting us to navigate through the different spheres of the smart city experience to discover its driving forces and effects.

This is why this book is divided into two parts. The first part, "The Smart City: What Transitions?", addresses, in six chapters, the issues of transitions related to digital protection and restraint, the dissemination of information, citizen inclusion, governance, transparency, the co-creation of public value and social acceptability. The second part, "The Smart City: What Way of Life for the Citizen?", analyzes, in eight chapters, the essential functions of "smart life": housing, food supply, consumption, mobility, health, culture, learning and tourism.

Each chapter addresses a dimension of the smart city by offering an original reading of the digital technological transformations at work and their concrete or potential effects on the daily lives of urban populations. Indeed, the starting point of each reflection is centered on the social needs and lifestyles that are directly affected or concerned by innovations. The approach developed then proposes to grasp the magnitude of the changes underway, tracing the design, but also the scope of innovations on social practices. The analysis thus emphasizes the individual and collective benefits as much as the challenges raised, or even the dangers incurred in connection with these transformations.

The richness of the work and the perspectives it provides are based both on the diversity of the subjects covered, which covers many facets of the smart city experience, but also on the multidisciplinarity of the points of view, which gives originality to the approaches. These perspectives are neither angelic nor pessimistic. The scientific foundations of the reflections guarantee an objective view of the situations examined and a critical distance that is useful for gauging the effective scope of the smart city and the challenges it raises. The book is therefore aimed at people who are curious to know more about everyday life in contemporary cities, whether they are scientists, students, elected officials or urban professionals.

I.2. References

Aït-Yahia, K.G., Ghidouche, F., N'Goala, G. (2019). Smart city of Algiers: Defining its context. In *Smart City Emergence*, Anthopoulos, L. (ed.). Elsevier, Amsterdam.

Deci, E.L. and Ryan, R.M. (1985). *Intrinsic Motivation and Self-Determination in Human Behaviour*. Plenum Press, New York.

N'Goala, G. and Collin-Lachaud, I. (2023). Incongruous trip to the Consumer Electronic Show (CES) in Las Vegas. *Décision Marketing*, 109(1), 133–139. doi: 10.3917/dm.109.0005.

PART 1

The Smart City: What Transitions?

Introduction to Part 1

"Smart city" strategies have been promoted around the world as an innovative technological response, of use to managers of urban areas, with the promise of unprecedented societal transformations for populations and territories. By focusing on digital technology, these strategies aim at both increased efficiency for many urban services, especially in networks (energy, transport, health, etc.), and at greater sustainability of urban systems, acting as much on supply as on use. But is this everything that can be said about the digital transition of cities? What are the other mechanisms at work, and how can we avoid the blind spots of an (overly) techno-centric approach?

By analyzing the advent of the smart city, the various contributions collected in this first part of the book draw attention to the other processes that constitute the smart city, while also highlighting the pitfalls to avoid or the rebound effects that may arise.

The implementation and scope of the smart city cannot be reduced to new digital infrastructures and the operational modalities of urban services. The ambition here is to deconstruct a monolithic approach to the smart city reduced to the advances of digital technology and services to recreate a deep multi-dimensional conception. Thus, the different chapters, in exposing what is behind the scenes of the smart city, confront it with its own demons. They reveal the whole set of mechanisms in operation, thus enriching an understanding of effects and collective action, from a perspective of the common good, ethics, and individual and collective freedoms.

Addressing all of the ins and outs of the smart city requires examining the actual scope of the transformations at work in improving urban management and the living conditions of populations. While the main goal of the emerging smart city is greater

---

Introduction written by Florence PAULHIAC SCHERRER.

efficiency of services, greater sustainability of urban systems and improved urban quality of life, it may partly fail, generating negative externalities. Furthermore, the political dimension of this goal and its impact on power relations, citizens and the collective interest or the common good should not be underestimated. In this context, we must highlight the environmental, social and political dimensions, which are just as constitutive of the transformations of this city as digital technology.

The understanding of the smart city is enriched here by critical perspectives in two ways. First of all, these chapters clearly emphasize that the processes of transformation linked to the smart city are risky and involve serious contradictions or paradoxes. For example, in trying to promote urban sustainability, the digital transition must also limit its ecological footprint and play the card of moderation or *digital sobriety*, if it wants to contribute to the ecological transition (Chapter 1). Another example shows that, in focusing on a broad deployment of digital infrastructures and services, the smart city must not generate new social inequalities, but, on the contrary, promote digital inclusion (Chapter 2). However, the conditions of access to services are inequitable and subject to nonuse. They may also incite resistance or mistrust. These different challenges are all the greater because companies, public organizations and local populations have different visions of the issues at stake or the role that they can play.

Furthermore, several chapters also show that the smart city has reshuffled many cards related to the roles of urban, private and public actors, including citizen-users, and urban power. Within the smart city, governance has changed with new players entering the arena. The impacts are potentially major, disrupting power relations, the conception of the public good and of citizenship. Several approaches underline that urban governance offers a more open reconfiguration of the arenas of negotiation and power. Thus, the figure of the user-citizen may make a formal contribution to the recomposition of decision-making processes, thus suggesting that participatory democracy is a key element of several smart cities, and that these latter are likely to opt for new public values (Chapter 3). However, the social contract and ethics of the smart city are regularly debated, particularly from the perspective of digital transparency relating to the collection and use of personal data (Chapter 4). This issue is not limited to formalizing explicit rules for this use. It involves the design of relationships of trust between companies and their customers, and even mechanisms for control and appropriation by the latter of the digital worlds that are imposed upon them.

Thus, within the smart city, the exercise of citizenship supposes knowing, understanding and acting in a perspective of collective interest through forms of empowerment, digital education (Chapter 4) and participatory mechanisms. Active citizen participation may facilitate the social acceptability of innovation (Chapter 5) by proposing to mobilize and disseminate precise, accessible and impartial information, and thus open a structured dialog. The role of those in positions of power is crucial in ensuring its effectiveness and enabling collective decision-making on what is feasible, desirable and fair. From this perspective, the information disseminated is therefore a key element. However, the contemporary media universe struggling with the revolutionary performances of artificial intelligence offers the risk of false information. Faced with the virus of disinformation, it becomes necessary to develop procedures for fact-checking, but also for inoculation and encouragement of media literacy (Chapter 6).

1

Protection

1.1. Introduction: a sustainable and sober city

Since their emergence in the 1980s, smart cities have increasingly developed in both France and internationally. Not only do they constitute an important lever of attractiveness for territories, but they also promise better management of cities and their resources. Today, five cities in France are considered "smart", according to the 2021 smart city index, with Bordeaux, Lyon, Lille, Paris, and Marseille at the top of the rankings.

Smart cities use smart connected devices to efficiently manage services (surveillance, mobility, health, administration, entertainment, delivery by autonomous intelligent robot, etc.) and city resources, including water, energy and waste management services (Lenouvel and Poiroux 2018).

In Lyon, for example, the metropolis has invested more than 340 million euros in smart city projects, such as smart grids to better preserve energy resources, or designing new neighborhoods better adapted to climate change (less concrete and more trees and greenery).

The goal is to ensure urban development with a strong ecological dimension. Paris also uses digital infrastructures to collect real-time data on energy, water and gas consumption in the building sector, enabling the city to make decisions more suitable for users and thus to reduce a waste of resources (Cisco). In Montpellier, smart initiatives are now making it possible to educate and support citizens in their ecological transition, in particular using digital systems for real-time reporting of consumption data (water, gas, electricity, mobility, waste, etc.) and customizable coaching programs (E3D Environnement).

Chapter written by Inès KOLLI and Philippe COHARD.

Digital technology offers several perspectives of development for cities and their inhabitants with a view to the ecological transition (Ahn et al. 2016; Cerri and Terzi 2016). Nevertheless, the impact of digital technology on the environment is now raising major questions. Indeed, the strong growth in the number of connected devices, as well as the overconsumption of digital services, sometimes called "digital obesity" or "infobesity", could lead to counterproductive effects for the environment (greenhouse gas emissions, overexploitation of rare earths necessary for manufacturing these components and production of digital waste). The challenge for stakeholders in the development of smart cities is therefore to find the balance between innovation through digital technologies and sustainability. To achieve this, sobriety is seen as a promising perspective for the future of cities. The challenge is to initiate the transformation toward smart cities in a more sustainable way and with a certain moderation, in accordance with the principle of sobriety for cities and territories.

In this chapter, we propose to explore the notions of sobriety for cities from the perspective of sustainability and that of smart cities. A study based on semi-structured interviews with experts and citizens of smart cities will allow us to answer several questions: is the development of digital technology in sectors of the city a lever or, on the contrary, an obstacle to environmental sobriety? Does the smart city allow effective management of the problems of overconsumption and overproduction of waste? How can we reconcile both the "digital" and "sober" approaches to the city?

1.2. Digital city and sober city: reconcilable issues

In the literature of management science, the concept of the smart city mainly refers to ways in which digital technologies might improve the functioning of cities, their efficiency, their competitiveness and provide new ways to resolve several dysfunctions (Batty et al. 2012). However, although digital technology is a crucial and inseparable aspect of this (Bibri and Krogstie 2017), the smart city is not only a digital city. For example, Cocchia (2014) emphasizes the need to broaden the focus of smart cities to make them more balanced and inclusive, stressing the importance of considering the different environmental, social and economic aspects of the city. Indeed, according to Albino et al. (2015), the concept of the smart city is linked in particular to that of sustainability, because it is now important to deal with massive urbanization which leads to heavy environmental and economic constraints. Among authors who have discussed the smart city, Giffinger et al. (2007) are some of the pioneers who proposed a typology of smart cities composed of six pillars, in which the *smart environment* dimension represents an important strategic axis for the development of cities (see Figure 1.1).

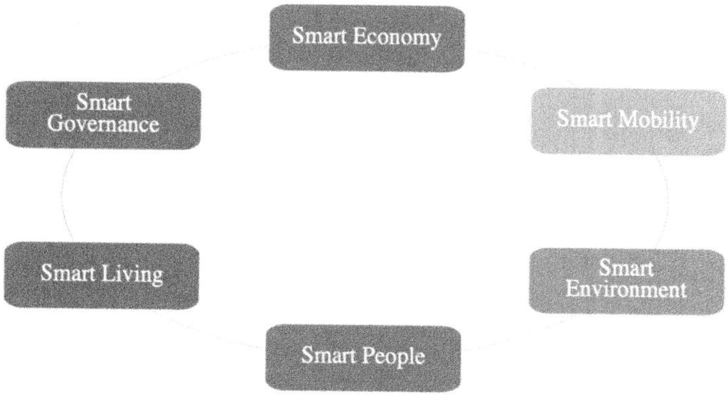

Figure 1.1. *Six pillars of the smart city (Giffinger et al. 2007). For a color version of this figure, see www.iste.co.uk/ngoala/smartcity.zip*

In this perspective of sustainable urban development, more and more solutions based on the Internet of Things and Big Data are being offered to the market by companies and private groups, such as Cisco, IBM and Veolia, for the use of public stakeholders.

NOTE.– The Internet of Things (IoT) links operational technologies, such as sensors and actuators, with information technologies, such as data processing and networking (Fagan et al. 2021, p. 1). Big Data, meanwhile, is a massive data set originally presented according to the 3Vs (Laney 2001): volume, variety and velocity.

Thanks to the implementation of IoT, the installation of sensor devices and collection networks across the territory, communities can measure air quality in real time, optimize energy networks, detect leaks in water networks, or even intelligently collect the city's waste. Indeed, sensors installed in city infrastructures allow the collection of large masses of data which may then be processed by statistical methods to create new knowledge (Yadav et al. 2017). All of these initiatives are not only economically beneficial, but also have the potential to reduce the city's carbon footprint.

NOTE.– The title of this chapter, "Protection", applies more to a vision oriented toward sustainable development, but it seems essential, in connection with the IoT and Big Data, to recall the need to consider the security of information systems and associated technologies. To go further on these questions, we invite you to consult the websites of the National Agency for the Security of Information Systems (ANSSI) or the European Union Agency for Cybersecurity (ENISA).

Furthermore, in parallel with the appeal of the concept of smart cities, the notion of sobriety for cities is gradually developing in France, encouraged by public policies in matters of eco-responsibility. Indeed, faced with the increase in the world population living in urban areas and the diversification of consumption needs, massive effects on resource extraction have been noted. Public and national bodies, particularly ADEME, have launched numerous initiatives to combat the waste of resources. This is particularly the case with the "zero trash, zero waste" approach proposed by certain local authorities, such as Nantes Metropolis in 2014 or Greater Rodez in 2016. Stakeholders are now seeking to encourage more sober initiatives, which can be seen in the framework of the circular economy.

In addition, the national legislative framework is currently experiencing remarkable developments aimed at initiating ecological change and facilitating the emergence of new systems to limit waste production and preserve natural resources, biodiversity and the climate, for example, (i) law No. 2021-1104 of August 22, 2021 on combating climate change and strengthening resilience to its effects, defined within the framework of the citizens' convention for the climate; (ii) the law relating to the energy transition for green growth of August 2015; and (iii) the anti-waste law for a circular economy (AGEC), promulgated in February 2020. According to the AGEC law, the circular economy (see Figure 1.2) calls for sober and responsible consumption of natural resources and raw materials, as well as, in order of priority, prevention of the production of waste, notably through the reuse of products, recycling, or, failing that, recovery of waste.

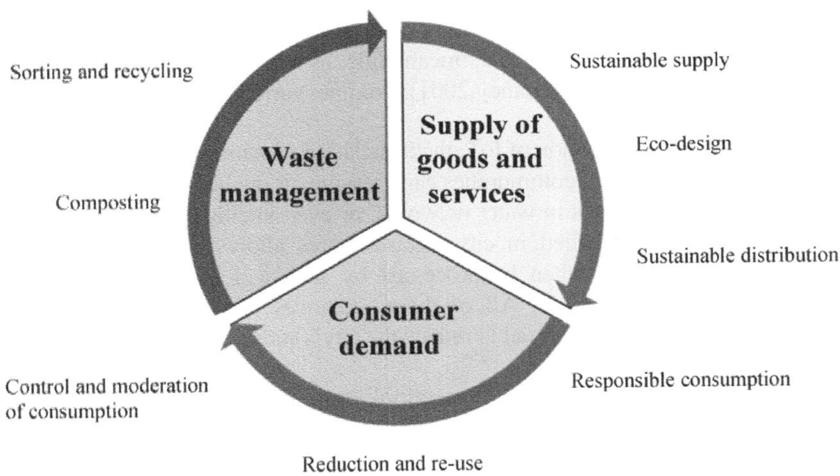

Figure 1.2. *Domains of action and pillars of the circular economy inspired by ADEME*

In the same prospect, there is an increasing number of French public funds that support entrepreneurial initiatives and accelerate the dynamics of the ecological transition. The state has entrusted ADEME with the mission of supporting on-the-ground implementation of waste policy, using the waste fund amounting to 193 million euros in 2016. In addition, funds amounting to 140 million euros have also been allocated over the period 2021–2022 to support investments dedicated to incorporating recycled materials into products as part of the recovery plan included in AGEC. Also, to facilitate the shift toward a circular economy, the Ministry of Ecological Transition has enacted several measures aimed at individuals. These public policies seek to raise awareness of environmental issues and encourage them to adopt more sober practices (zero waste practices, consumption control).

Sobriety in the construction sector

Construction activities produce more than 220 million tons of waste per year, including 185 million from public works and 42 million from building (ADEME 2020). According to ADEME, this volume represents 14% of the waste currently produced in France (ADEME 2020). Faced with this observation, sobriety in the construction sector aims to minimize the impact on the environment and optimize the consumption of resources, while maintaining the quality of construction. For example, thanks to its circular economy and zero waste policy, the city of Montpellier invites us to rethink how the construction sector operates by encouraging practices of recovery and reuse of construction materials. In this context, the MRBC building materials reuse center has developed the first platform for purchasing and reselling reusable materials (beams, pillars, slabs, joinery, pipes, etc.), with the goal of reducing the quantity of waste produced.

Sobriety in the catering sector

Another example of sobriety in the home food delivery sector is the company LoopEat. This company offers a reuse provision based on the deposit system. Meals are delivered in returnable containers and customers are rewarded for using their own containers. The aim is to ensure sobriety in packaging by eliminating single-use plastic.

Box 1.1. *Example of sobriety initiatives in France*

In the literature, sobriety is defined as reserve, restraint or moderation in consumption (Cézard and Mourad 2019). It is manifested in several practices: reduction, reuse, recycling, etc. (De Lanauze and Siadou-Martin 2013; Rumpala 2018; Van Dormalen and Kleesattel 2018; Guillard 2019; Săplăcan and Márton 2019; Defontaine 2020). Sobriety consists of questioning our real consumption needs and satisfying them while limiting their impact on the environment. It must lead to changes in methods of production and consumption (more use of bulk sales, re-examining the functions of packaging, etc.), and, more generally, lifestyles on an individual and collective scale. Applied to the context of cities and territories, sobriety consists of questioning the real needs of the population and in designing the city while respecting the principle of controlling energetic, material and digital production and consumption (Roudil 2021). This implies a significant reduction in the consumption of resources and a transformation of processes of production, city design (Roudil 2015), resource management and the lifestyles of individuals, following the principle of zero waste and circularity of resources. Today, this approach is seen as a more effective and promising approach to reduce environmental impact and accelerate the ecological transition. Box 1.1 presents some examples of sobriety initiatives in France.

Many academic works emphasize that technological progress, and especially digital progress, is a source of many advantages for sustainable development (Ahn et al. 2016; Cerri and Terzi 2016). Ferreboeuf (2019) reports that digital technology is seen as a solution for optimizing energy consumption in several sectors, particularly through more efficient use of energy resources (smart grids), the transport network (smart mobility), e-commerce digital services or even agriculture (smart farming and smart water). From the point of view of users, several green initiatives are offered on the market, such as the Linky meter, which makes it possible to monitor electricity consumption in real time and take appropriate measures to optimize energy efficiency.

However, despite these numerous initiatives, the smart city has not been fully reconciled with the sober city. For many authors, digital technology is paradoxically following a worrying ecological trajectory. While the smart city is based on integrating digital technologies into city services to optimize their management, many argue about the harmful impact of technologies on the environment and thus call for digital sobriety (Bordage 2019a; Elgaaied-Gambier et al. 2020), or even "digital degrowth" (Ferreboeuf 2019; Marquet et al. 2019). The French association The Shift Project reports that the growth of digital systems is now unsustainable, thus implying significant issues in terms of waste and excess consumption of electricity (an increase of more than 9% in 2020). The studies of Bordage (2019a) or Marquet et al. (2019) offer an overview of the major impacts of digital technology. In this sense, the production phase of digital equipment involves heavy consumption of the resources needed to manufacture its components (copper, platinum, gold, gallium, indium, germanium, etc.). This production is in fact a source of multiple forms of pollution of soils, rivers and air occurring at the different stages of extraction, chemical treatment of materials and removal of impurities. In addition, the process of manufacturing and assembly of the terminals involves high consumption of electricity. This consumption represented around 44% in 2020 of the total energy footprint of digital technology and the resulting CO_2 emissions (Bordage 2019b; Ferreboeuf 2019). In the disposal phase, end-of-life management (recycling, repair) of digital equipment represents a major challenge for societies. In France, around 995,000 tons of electrical and electronic equipment waste were collected in 2021 for nearly 1.4 billion items of equipment placed on the market (ADEME 2022).

Placing the objectives of the circular economy on the agenda invites us to rethink the trajectory of development of smart cities and suggests that its stakeholders should anticipate societal expectations and regulatory developments in favor of more sustainable and more sober development of cities. Several questions are thus raised: to protect the environment, how can we reconcile the technological progress of smart cities with the challenge of sobriety, of combating waste of resources, of reducing overconsumption and the overproduction of waste? What is the place and the role of digital technology in achieving the goal of sobriety?

1.3. A qualitative study for better understanding the development of smart cities in view of sobriety

To answer our question, this research is based on semi-directive interviews conducted with the following three types of smart city stakeholders: private companies, public institutions and citizens. In total, 24 semi-directive interviews were conducted with two energy and environment companies, a cooperative firm ("SCOP"), two digital service providers, a smart technology provider, two metropolises, a smart cities consulting firm, a smart cities association and 12 citizens (see section 1.6). This group was formed to have a fairly broad perspective on perceptions of the smart city and its challenges from different types of stakeholders. The citizens interviewed for this study are not employed in a role directly related to the smart city.

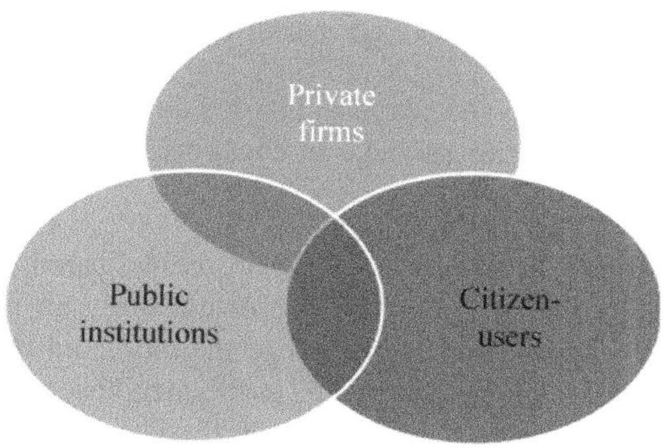

Figure 1.3. *Smart city stakeholders interviewed in the qualitative study*

1.3.1. *A techno-centric approach to the smart city*

Analysis of the results shows that several experts support a techno-centric vision of the smart city and its major role in the development of more ecological cities. One expert tells us: "The smart city refers to the question of the digitalization of infrastructure, when you have a building connected with sensors, you understand better what is happening inside, when a road is going to be connected: less public lighting, you optimize energy" (stakeholder no. 9). The experts stress the importance of digital progress in addressing current urban issues (energy and resource management, road traffic management, security, etc.): "In fact, the smart city and digital technology are becoming a necessity today for doing many things [...], for example, in terms of optimizing travel, energy optimization, digital technology is already present everywhere" (stakeholder no. 7). Some argue that digital technology plays a major role in sensitizing citizens–users of the city to environmental issues. One expert reports: "With our databases, we managed to develop an application saying: well, you are higher than the average (energy consumption of a household); you are average; you are below, be careful" (stakeholder no. 1).

The experts seem to be aligned with a techno-centric vision of the smart city. However, to reconcile the two visions of the digital city and the sober city, and to better protect the environment, it seems important to collectively re-discuss real digital needs and to use information technologies in accordance with the objectives of sobriety. In this context, several actions seem necessary for the development of smart cities, notably control of digital energy consumption (storage in green data centers, energy management for IT assets, etc.) and reduction of digital waste through the fight against planned obsolescence (AGEC law 2021), the use of standardization and reparability systems for connected devices to avoid waste.

1.3.2. *Different visions between experts and citizens of the smart city*

The results of the study show the need to involve citizens in collective action and emphasize the difficulty that experts have in engaging citizen–users in practices of sobriety. One expert tells us:

> Afterwards, where it becomes much more complicated is when we finally arrive at common ground, to the extent that I am ready to get involved in the collective management of the city. And there, we can find a lot of things, we can say: well, in this neighborhood, there are going to be redevelopment projects. What makes it so that, at any given moment, the 'smart city', and everything being done today with digital technology, can allow me to get interested in what is happening? So, to become interested in getting involved in it? And I haven't found a trigger for that yet! (stakeholder no. 2).

These elements are in agreement with a previously conducted study on the place of the citizen (Kolli et al. 2020).

Furthermore, analysis of the interviews conducted with citizens highlights the reluctance of certain individuals to get involved in preserving the environment due to the significant efforts required, especially in the smart environment and smart mobility areas. This reluctance is explained by the real or perceived sacrifices and efforts that are required. For example, one respondent states: "It's a pain to do [sorting waste]. Turning down the heating is also a pain. So, in fact, it's just burdensome actions" (stakeholder no. 19). Others, on the contrary, are ready to commit to the environmental cause without necessarily wanting to be part of the smart city system: "I don't need an app on my smartphone to further convince me that the planet is not doing well [ironic laugh], I'm already more than convinced. It's like last time, a friend takes the phone, she looks at it and she tells me: it's raining. Well, I say: yes, it's raining, obviously [derisively]! She's not an idiot, but honestly, what... Please, what?" (stakeholder no. 22).

Citizens show an ambivalent attitude toward the use of new connected devices for environmental protection (energy, waste, water and pollution). Respondents cite the notion of "novelty" as a factor that may encourage the use of new environmental technologies; however, their availability may not be enough to change environmental practices in the long term, as one respondent points out: "I don't know. In the short term, yes, because I do like the technology aspect and the app aspect, so I imagine that if someone tells me: hey, there's such-and-such an application, I'll download it and try it, just to see. But I'm not sure if it will last in the long term, you'd have to try to find out" (stakeholder no. 24). The use of "new" digital devices appears to be a lever to encourage citizen involvement in the development of a more sober city, but perceived usefulness plays an important role.

Indeed, for citizens, usefulness as perceived individually remains the main determinant favoring the use of environmental connected devices. This seems to align with the Technology Acceptance Model (TAM) (Davis 1989). In the field of energy, new systems for measuring rates of energy consumption and providing real-time notifications might have an influence on adoption of eco-responsible practices where this results in personal gain (saving money): "These feedback systems for energy consumption are interesting! Especially if you spend a lot, if you have a big house, I think that would be important" (stakeholder no. 15).

However, when a new connected device aims to mobilize citizens toward preserving the collective interest, the reactions of respondents remain relatively negative. For example, for a device for real-time measurement of pollution rates in in the city, one respondent tells us: "Uh [hesitation]... Yes, it's good to know how much pollution there is, but after that, what's the point in the end? It's not like checking the weather forecast because I want to know what the weather's like tomorrow so I know what to wear!" (stakeholder no. 13).

Thus, while many citizens are ready to adopt new technologies to change their practices, some believe that the use of technology will not be enough to encourage citizen involvement in community life, or even to encourage them to adopt more sober practices aimed at the well-being of the community and the environment. To successfully engage individuals in ecological practices, smart green technologies must integrate prior reflections on the psychological mechanisms underlying individual behavior into their design. More precisely, the perceived mechanisms of effort and sacrifice associated with ecological practices must be considered.

1.4. Sustainable and sober city: key points

To better protect the environment, it is now necessary to link the approaches of sobriety and digital progress. Smart city project managers should anticipate the evolution of society's concerns by adopting a more global vision of the city's smart innovations and by involving citizens more in the development process of sober cities.

From the point of view of providers of urban environmental services (water cycle management, city waste and mobility), although the literature shows that digital technology (Big Data, artificial intelligence algorithms, smart chips and sensors, etc.) offers the advantage of data circularity and consumption control due to data collected in real time, alert systems and decision-making support (Jabbour et al. 2017; Patwa et al. 2021), this does not exempt them from analysis of their effects on the environment. From a perspective of sobriety, stakeholders should account, in their strategic decision-making processes, for the carbon dimension of smart technologies, and prioritization of devices that allow compensation for the effect of the technology itself in terms of greenhouse gas emissions. These elements might also be an incentive for technology suppliers from the design and use phase to the disposal phase of the device.

From the point of view of local authorities and public managers, evaluation of systems linked to technologies and their uses appears particularly necessary. The following articles may provide some leads for conducting such studies around the unified theory of technology acceptance and use (Lebrument et al. 2021), as well as the value of information systems (Cohard 2020). Sustainability criteria might also be considered in the choice of solutions implemented for the smart city.

From the point of view of providers of digital objects intended for users (smartphones, apps, etc.), a strategy of sobriety would consist of offering repairable, modular, standardized and more durable products. Companies would combat unsustainable practices and promote strategies for reusing materials, manufacturing low-energy consumption products and offering devices made from recyclable and recycled materials.

Furthermore, in a sobriety approach, the challenge would be to massively mobilize the population with the ecological objective of a thoughtful use of technologies. Involving the citizen–user in the transition toward a more sober city requires a change in daily practices. Notably, previous research (Kolli et al. 2020) showed several levels of involvement expected by citizens: for example, the use of platforms for buying and reselling second-hand products, the use of information and decision-making support applications as well as notification and support systems for changing individuals' consumption behavior.

Sobriety from the consumer's point of view is also reflected in practices of reduction and control of consumption of digital services. Indeed, Ferreboeuf (2019) discusses two individual practices that have a harmful impact on the environment. On the one hand, a high replacement frequency for terminals (smartphones, computers and other connected devices). For example, in France, a smartphone is replaced after an average of 2–2.5 years (*The Shift Project* report 2018, for digital sobriety) and 88% are changed while they are still working (ADEME 2018).

On the other hand, there is a strong growth in the number of digital devices bought and used by individuals. According to a study published by Bordage (2019a), in 2018, 15 billion computers, game consoles, smartphones and other digital devices were recorded. These devices are used by more than four billion individuals who connect every day, the equivalent of 30 billion hours of use[1]. In addition, the development of new digital uses, including free streaming videos, viewing images in HD and UltraHD quality, cloud storage services, e-commerce, social networks and online digital games generate a strong increase in traffic on the networks. The studies of El-Gaaied-Gambier et al. (2020) list several practices of digital sobriety to adopt from a consumer perspective. They highlight the positive impact on the environment of several practices, such as limiting the number of emails sent and reductions in the number of people copied into emails and in cloud data storage.

Finally, in addition to identifying sobriety practices to adopt on a daily basis, our qualitative research highlights the difficulty of really engaging consumers in this objective of sobriety. Understanding the psychological mechanisms underlying human behavior and identifying the obstacles associated with sobriety, especially the perceived efforts and sacrifices (Kolli 2022), should push managers of behavioral change to deploy appropriate techniques of education and persuasion to minimize psychological obstacles and thus hope for positive effects over the long term (Okur and Saricam 2019).

1.5. Conclusion

Between digital progress and sobriety challenges, what trajectory might the smart city take? The smart city, as it is developing today, is based mainly on a

1 See: www.greenit.fr.

techno-centric approach that aligns imperfectly with current issues of sobriety. The lack of questioning and collective discussion between digital designers and those responsible for change in terms of environmental sobriety may generate divergences between stakeholders, but also confusion in the minds of end users-consumers. This is why it is interesting to interrogate the place of digital technology in a city that advocates sobriety. Stakeholders would do well to rethink business models in the light of sobriety approaches, beginning from the prior design of smart devices. Sobriety would also involve considering the psychological mechanisms underlying individual behavior (psychological levers and obstacles) in the processes of designing connected devices and encouraging users.

1.6. Appendix: profiles of respondents to the qualitative study

Stakeholder no.		Gender	Age	Profession
Smart city experts	1	F	/	Representative of an energy group in the Occitania region
	2	M	57	Consultant in innovation and smart territories
	3	F	47	Delegate of the confederation of cooperative firms
	4	M	/	Director of energy supply
	5	M	/	Innovation director – metropolitan
	6	F	37	Entrepreneur – director of technological solution development
	7	M	46	Director of digital acceleration of countries
	8	M	39	President of smart city association
	9	F	/	Smart city project manager – metropolitan
	10	M	/	Industry and energy marketing director
Citizens	11	M	28	Young graduate
	12	F	27	PhD student
	13	F	30	Student
	14	F	37	Optician
	15	M	29	Medical laboratory analyst
	16	F	65	Retired
	17	M	46	Entrepreneur
	18	M	30	PhD student
	19	F	25	Project manager
	20	M	38	Lawyer
	21	M	40	Unemployed
	22	M	61	Retired
	23	F	28	Employee
	24	F	29	Student

1.7. References

ADEME (2018). Les impacts du smartphone : un téléphone pas si smart pour l'environnement [Online]. Available at: https://presse.ademe.fr/wp-content/uploads/2017/09/guide-pratique-impacts-smartphone.pdf.

ADEME (2020). Déchets chiffres-clés. Édition 2020 [Online]. Available at: https://librairie.ademe.fr/ged/28/dechets-chiffrescles-edition2020-3-010692.pdf.

ADEME (2022). Équipements électriques et électroniques : données 2021. Annual Report [Online]. Available at: https://librairie.ademe.fr/ged/8215/Equipements-electriques-electroniques-EEE-donnees2021-rapport2022.pdf?Submit=.

Ahn, M., Kang, J., Hustvedt, G. (2016). A model of sustainable household technology acceptance. *International Journal of Consumer Studies*, 40(1), 83–91.

Albino, V., Berardi, U., Dangelico, R.M. (2015). Smart cities: Definitions, dimensions, performance, and initiatives. *Journal of Urban Technology*, 22(1), 3–21.

Batty, M., Axhausen, K.W., Giannotti, F., Pozdnoukhov, A., Bazzani, A., Wachowicz, M., Portugali, Y. (2012). Smart cities of the future. *The European Physical Journal Special Topics*, 214(1), 481–518.

Bibri, S.E. and Krogstie, J. (2017). Smart sustainable cities of the future: An extensive interdisciplinary literature review. *Sustainable Cities and Society*, 31, 183–212.

Bordage, F. (2019a). *Sobriété numérique : les clés pour agir*. Buchet, Chastel.

Bordage, F. (2019b). *Empreinte environnementale du numérique mondial*. Greenit.fr, Paris.

Cerri, D. and Terzi, S. (2016). Proposal of a toolset for the improvement of industrial systems' lifecycle sustainability through the utilization of ICT technologies. *Computers in Industry*, 81, 47–54.

Cézard, F. and Mourad, M. (2019). Panorama sur la notion de sobriété – définitions, mises en œuvre, enjeux. Final Report, ADEME [Online]. Available at: https://librairie.ademe.fr/ged/491/rapport-etat-lieux-notion-sobriete-2019.pdf?modal=false.

Cocchia, A. (2014). Smart and digital city: A systematic literature review. In *Smart City*, Dameri, R.P. and Rosenthal-Sabroux, C. (eds). Springer, Cham, 13–43. doi: 10.1007/978-3-319-06160-3_2.

Cohard, P. (2020). Information systems values: A study of the intranet in three French higher education institutions. *Electronic Journal of Information Systems Evaluation*, 23(1), 150–167.

Davis, F., Bagozzi, R., Warshaw, P. (1989). User acceptance of computer technology: A comparison of two theoretical models. *Management Science*, 35(8), 982–1003.

De Lanauze, G.S. and Siadou-Martin, B. (2013). Deconsumption practices and motivations: A theory of value approach. *Revue Française de Gestion*, 230(1), 55–73.

Defontaine, G. (2020). La sobriété : indispensable à une transition réussie ? *Regards croisés sur l'économie*, 26(1), 153–160.

Elgaaied-Gambier, L., Bertrandias, L., Bernard, Y. (2020). Cutting the Internet's environmental footprint: An analysis of consumers' self-attribution of responsibility. *Journal of Interactive Marketing*, 50(1), 120–135.

Fagan, M., Marron, J., Brady, K., Cuthill, B., Megas, K., Herold, R. (2021). IoT device cybersecurity guidance for the federal government: Establishing IoT device cybersecurity requirements. NIST Special Publication 800-213, Gaithersburg, MD. doi: 10.6028/NIST.SP.800-213.

Ferreboeuf, H. (2019). Pour une sobriété numérique. *Futuribles*, 2, 15–31.

Giffinger, R., Fertner, C., Kramar, H., Kalasek, R., Pichler-Milanovic, N., Meijers, E. (2007). *Smart Cities – Ranking of European Medium-sized Cities*. Vienna University of Technology, Vienna [Online]. Available at: http://www.smart-cities.eu.

Guillard, V. (2019). *Du gaspillage à la sobriété : avoir moins et vivre mieux ?* De Boeck Supérieur, Brussels.

Jabbour, C.J.C., de Sousa Jabbour, A.B.L., Sarkis, J., Godinho Filho, M. (2019). Unlocking the circular economy through new business models based on large-scale data: An integrative framework and research agenda. *Technological Forecasting and Social Change*, 144, 546–552.

Kolli, I. (2022). Le sacrifice perçu pour l'environnement : conceptualisation, antécédents et impacts sur l'adoption des pratiques de zéro-déchet et de déconsommation. PhD Thesis, Université de Montpellier, Montpellier.

Kolli, I., Laurent, S., Cohard, P., N'Goala, G. (2020). *Placer le citoyen au cœur de la Smart City. Management des technologies organisationnelles*. Presses des Mines, Paris.

Laney, D. (2001). 3D data management: Controlling data volume, velocity and variety. Research Note, Meta Group.

Lebrument, N., Zumbo-Lebrument, C., Rochette, C. (2021). L'acceptation des applications mobiles de MaaS : une application de l'UTAUT2 au contexte des villes intelligentes françaises. *Systèmes d'information & management*, 26, 19–54.

Lenouvel, V. and Poiroux, C. (2018). Typologie de modes d'adoption des technologies alternatives pour la ville sobre. In *Villes Sobres*, Lorrain, D., Halpern, C., Chevauché, C. (eds). Presses de Sciences Po, Paris.

Marquet, K., Combaz, J., Berthoud, F. (2019). Introduction aux impacts environnementaux du numérique. *Bulletin de la Société Informatique de France*, 13, 85–97.

Okur, N. and Saricam, C. (2019). The impact of knowledge on consumer behaviour towards sustainable apparel consumption. In *Consumer Behaviour and Sustainable Fashion Consumption*, Muthu, S.S. (ed.). Springer, Berlin.

Patwa, N., Sivarajah, U., Seetharaman, A., Sarkar, S., Maiti, K., Hingorani, K. (2021). Towards a circular economy: An emerging economies context. *Journal of Business Research*, 122, 725–735.

Roudil, N. (2015). La ville durable à l'épreuve de la sobriété. Le citadin entre injonction à "bien habiter" et normalisation des conduites en milieu urbain. In *Systèmes énergétiques renouvelables en France et en Allemagne, synergies et divergence*, Hamman, P., Christen, G., Jehling, M. (eds). Orizons, Paris.

Roudil, N. (2021). La fabrique de la "ville sobre" en France : les sciences humaines et sociales au risque de contribuer au processus de normalisation des conduites et d'incitation au changement. *Sociologies*. doi: 10.4000/sociologies.17902.

Rumpala, Y. (2018). Quelle place pour une "sobriété heureuse" ou un "hédonisme de la modération" dans un monde de consommateur ? *L'Homme et la Société*, 3, 223–248.

Săplăcan, Z. and Márton, B. (2019). Determinants of adopting a zero-waste consumer lifestyle. *Regional and Business Studies*, 11(2), 25–39.

Van Dormalen, L. and Kleesattel, S. (2018). The consumer journey to adapt a zero-waste lifestyle: A transtheoretical approach. Master's Thesis, Jönköping University [Online]. Available at: https://www.diva-portal.org/smash/get/diva2:1210804/FULLTEXT01.pdf.

Yadav, P., Hasan, S., Ojo, A., Curry, E. (2017). The role of open data in driving sustainable mobility in nine smart cities. In *Proceedings of the 25th European Conference on Information Systems*, Guimarães, 1248–1263.

2

Including People

2.1. Introduction: a digital city for all citizens

The digital transformation of cities, often referred to as the smart city concept, raises profound social and managerial questions. By optimizing a city's functioning through the use of digital technology, the smart city aims to create a living environment for all citizens that fosters their ability to act and participate. For Ratti (2017), the key to urban success lies in building together with all stakeholders in the city. *The smart city's collaborative approach places citizens and their relationship with digital technology at the center of discussions*: how can we design a "digital" way of living together that includes all citizens, including those most estranged from digital technology? As Ghorra-Gobin (2018) suggests, does the "optimization of our urban services (thanks to sensors, connected objects and the Internet) really allow us to re-enchant democracy"?

The massive deployment of digital technology and its uses in our daily lives (administrative tasks, customer experience, remote working, consumption, etc.) continues to grow (see Box 2.1). The Covid-19 health crisis further accelerated the digitization of interactions between organizations and users. In 2020, we observed an acceleration in digital adoption equivalent to three additional years compared to the average rate between 2017 and 2019.

Nevertheless, given the many paradoxes generated by digital technology and its uses, inequalities may appear within the population: "thirty-five percent of French people experience at least one form of difficulty preventing them from fully using digital tools and the Internet". The ninth edition of the barometer of French confidence in digital technology also observes that more than one in two French people think that using the Internet is risky. Faced with these digital divides, various

---

Chapter written by Léa CAUCHARD.

stakeholders in the city (institutions, associations, businesses, education stakeholders, etc.) engage their social responsibility so as not to exclude those citizens most estranged from digital technology. They thus implement actions of digital inclusion which aim to promote access to digital technology and training for those citizens deemed to be "excluded" (see Table 2.1). However, there remain considerable disparities within the population, leading to the question: are we using the right definition of digital inclusion? How to define digital inclusion? How can urban stakeholders adopt smarter approaches?

In 2020, digital technology represented 6% of French GDP, or an estimated share of 150 billion euros.

In 2022, 83% of the 250 most frequently used procedures were paperless.

Since 2009, the digital sector has recorded growth of "more than 20%". More specifically:

– 58% of mayors of municipalities with more than 20,000 inhabitants have a digital strategy.

– Nine out of 10 communities have begun their digital transformation to improve the quality of life of citizens through digital technology, and 69% of them have made it a priority.

Digital technology is a high-demand sector: 175,000 net jobs have been created in France since 2009 and 232,000 recruitments are planned by 2027.

Following the establishment of DINUM (Interministerial Digital Directorate) in 2019, 908 million euros were invested as part of the France Relance Plan, with 250 million euros specifically allocated to digital inclusion. According to the common framework for digital uses of the Regulatory Authority for Electronic Communications, Postal Services and Press Distribution (Arcep) and the Regulatory Authority for Audiovisual and Digital Communication (Arcom, merger of the CSA and HADOPI), published on March 16, 2022: 78% of SME/VSE managers see digital technology as a source of profit for their organization. In fact, "by 2023, 50% of G2000 companies will devote half of their new spending on technology hardware and connectivity to modernizing and redesigning in-person experiences for customers and employees on their own premises".

Box 2.1. *Deployment of digital technology in France, some figures*

This chapter provides insights into both the understanding of digital inequalities and the process of citizen inclusion in a smart environment. It seeks to define the role that digital technology must play in this process of individual inclusion. How can urban stakeholders establish themselves in smarter approaches? In other words, with approaches that are both digital and inclusive, promoting the well-being and participation of all.

2.2. Understanding "digital inequalities": citizens' relationship to digital technology

2.2.1. *The phenomenon of the digital divide*

The digital divide relates to inequalities in participation in the digital society (Yu et al. 2018). This phenomenon illustrates the gaps existing within the population between individuals comfortable with technology – the included – and individuals estranged from technology – the excluded. This divide can be represented as a boundary that can be crossed in one direction – from exclusion to inclusion, as well as in the other – from inclusion to exclusion (Vodoz 2010). Kiyindou thus defines the digital divide as "the gap between those who use the potential of information and communication technologies for their personal or professional fulfillment and those who are not in a position to use them due to lack of access to equipment or skills" (Kiyindou 2007, p. 1).

Even though digital infrastructures are developing and issues with digital access are decreasing year on year, there are still significant disparities in access in France. Furthermore, although access represents an essential condition for moving toward digital inclusion, it is no longer possible to reduce the determinants of the digital divide to a binary dimension of access: having versus not having. Indeed, over many years, studies on the digital divide have identified numerous determinants, both extrinsic and intrinsic to individuals, that explain this phenomenon (Hargittai 2002; Hargittai and Hinnant 2008; Vodoz 2010; Yu et al. 2018). A new term may thus be preferred to "digital divide": differentiation of digital uses.

2.2.2. *Differentiation of digital uses*

Jauréguiberry (2003, 2005, 2012) observes that societal and behavioral developments have transformed individual perceptions of technology and its uses. He focuses on the voluntary nonuse of technology, which has hitherto been little represented in the literature. Voluntary nonuse is defined as nonuse that is desired by the individual, which is therefore not necessarily synonymous with discomfort and exclusion. In this case, by limiting their use of technology, the citizen seeks to limit their information overload: "We are moving from a problem of access to that of excess" (Jauréguiberry 2012). This study complements those of Cushman and Klecun (2006) and Kellner et al. (2010) who define the three main reasons for digital nonuse: lack of need, lack of interest and relevance, and lack of motivation.

These factors create new divisions within the population: involuntary nonusers, who wish to be included in the digital society, and voluntary nonusers, who reject this ideological attachment (Granjon 2010). This distinction leads to consideration of "differentiations in digital usage" rather than inequalities in digital usage (Peter and Valkenburg 2006). Indeed, there is a varied typology of digital uses (Katz and Aspden 1998; Wyatt 2003; Livingstone and Helsper 2007) and "digital nonuse" is not always experienced negatively or synonymous with digital exclusion (Peter and Valkenburg 2006; Granjon 2009). It may reveal itself to be a citizen's state of mind – temporary or permanent – or even a desire to remain in control of individual choices (Dagenais 2012; Durand-Mégret et al. 2016). These differences of opinion may originate in the contradictory discourses around digital technology and its uses.

2.2.3. *Digital technology, vector of exclusion or vector of inclusion*

In an ultra-connected smart environment, two ideological hypotheses are opposed:

– "*lamentation*", which assumes that digital technology fuels inequalities within society;

– "*techno-centric optimism*", which credits digital technology with an emancipating role for citizens.

2.2.3.1. *"Lamentation": digital technology as a vector of exclusion and ill-being*

For Kenyon et al. (2002), the most excluded populations are those who experience the most difficulty accessing digital technology. For these authors, an individual's social participation strongly depends on their mastery of digital tools. These realities quickly lead to frustration among individuals (Meuter et al. 2000) to the extent of reducing their well-being and their sense of belonging to society.

This reduction in well-being can also occur in populations who use the Internet and social networks intensively/without self-control. Technological stress and excessive use of digital technology may lead to significant paradoxes of mood change, isolation or even depression (Haferkamp and Krämer 2011). Thus, depending on the situations and contexts, the quality of life of each individual, regardless of biographical background, may be diminished due to digital technology, even leading to a feeling of exclusion in the most extreme cases.

2.2.3.2. *"Techno-centric optimism": digital technology as a vector of inclusion and well-being*

The techno-centric approach is an optimistic approach that considers digital technology essential for achieving a better quality of life. In this sense, many authors consider digital technology to be necessary for greater social justice, greater autonomy and more assertive development of self and social capital, synonymous with performance and self-optimization. Aslam et al. (2020) indicate that technology and digital technology stimulate social inclusion and "significantly influence people's choices and decisions, which are shaped by the country's institutional structure" (Aslam et al. 2020, p. 336). For Wallcook and Morris (2017), digital technology reduces distance between individuals, in other words, the risks of social exclusion.

This forward march toward ever-greater integration of digital technology into pathways is often presented as "self-evident" and the guarantee of a win-win exchange between urban stakeholders and citizens, because:

> [It enables making] preexisting resources more quickly and easily accessible, but also and above all by the reconfiguration of social and political relations that the tool allows: greater transparency, reversal of the expert/client relationship, interactivity, authorizing disadvantaged communities to reclaim the communication space (Carter and Grieco 2000).

Beauchamps (2009) and Dennis et al. (2016) observe that social exclusion reduces the happiness and well-being of individuals, but that certain technologies, such as mobile technologies, may compensate for this sentiment.

Given these elements, clearly delimiting the boundary between digital inclusion and digital exclusion remains complex. The real risks of a digital divide within the population arise when differentiations in digital use give rise to a feeling of exclusion among individuals. More precisely, digital exclusion exists when digital uses lead to asymmetries between citizens as a result of a lack of participation for some, and not for others. This potential lack of participation excludes citizens from the various opportunities available within the city, such as when digital technology must be used to access their fundamental rights or carry out compulsory administrative procedures (declare their taxes, renew their identity papers, receive allowances, etc.).

Urban stakeholders must adopt a new approach that fosters closer interactions with citizens and their digital perceptions. How can we build more inclusive digital pathways in the smart city?

2.3. Digital inclusion at the level of the smart city: designing digital pathways for all citizens

2.3.1. *The inclusion process: action frameworks for urban stakeholders*

According to Wilson and Secker (2015), inclusion occurs on both individual and collective levels, reflecting the degree to which people feel accepted and integrated into different social contexts. This definition implies that the inclusion process is constructed on two levels.

– *On the collective level*: social inclusion is only possible when normative and environmental conditions create a favorable context. This is reflected in the presence of social norms, aid, trading conditions, rules, freedoms, etc., and by all of the opportunities rendered accessible to the citizen through the user pathways designed by urban actors.

– *On the individual level*: social inclusion is embodied in the feelings of citizens. Citizens will determine their social position based on evaluations of the different normative and environmental conditions and the quality of their social interactions: "Am I accepted?", "Am I considered?", "Do I have access?", "Can I thrive?", etc.

Gardou (2012) describes an inclusive society as one that adapts to provide a place for all, ensuring equal participation in social, cultural and community life. In a digital environment, smart city stakeholders must therefore support the digital transformation of the services offered by the city in order to build an environment and conditions favorable to spreading the feeling of digital inclusion among all citizens. Furthermore, according to Cohen's (2012) model, an inclusive society and the integration of technologies are among the factors that enable evaluation of the smart city. Participation of all citizens – smart citizens – must be ensured. However, some of them are still unable to contribute and benefit from the smart city, because they have neither the tools nor the knowledge and skills necessary to interact effectively with digital technology (and therefore to participate fully).

Indeed, the digital transformation of services leads to a double pressure on individuals: having the equipment and the digital skills to use it (Pinède 2018; Brotcorne et al. 2019). This dual requirement fuels the risks of exclusion between those citizens capable of meeting these requirements and others (Mazet 2019): citizens who are not capable of using digital services can neither contribute to the community, nor share their data, nor participate in the governance of services. Urban stakeholders must rethink and optimize the design of digital pathways to ensure accessibility for all.

2.3.2. *Inclusive design: theoretical foundations*

Fisk et al. (2018, p. 842) define inclusion in services as “an egalitarian system which provides customers with fair access to a service, fair treatment during a service, and a fair opportunity to leave a service”. They identify four pillars for constructing inclusive services, which we contextualize at the smart city level:

– offering citizens the opportunity to develop themselves and create value by facilitating access to and use of (digital) city services;

– allowing citizens to choose between several different service offerings (digital or nondigital, for example);

– providing all citizens with equitable access to essential services (basic needs and rights);

– fostering positive interactions during services “by welcoming and accommodating all [citizens] in a positive environment” (Fisk et al. 2018).

For Patrick and Hollenbeck (2021), inclusive design depends on three main actions: providing access, creating equitable experiences to promote citizen participation and creating seamless experiences to foster citizen autonomy. Inclusive design thus aims to “reduce the actual or perceived gap between the user and the object of design” (Patrick and Hollenbeck 2021, p. 1).

More precisely, the authors draw on theories of evaluation and social inclusion to conceptualize individual responses to inclusive design. Their *DARE* framework (see Figure 2.1) breaks down an individual’s cognitive mechanisms during a pathway: the pathway sends signs of inclusion/exclusion to the individual who evaluates them before determining the existence (or nonexistence) of a gap between their identity and the characteristics of the pathway. Depending on the degree of perceived discrepancy, their emotional responses and lived experience will be more or less positive/negative.

Fisk (2018) and Patrick and Hollenbeck (2021) thus offer initial avenues for reflection for urban stakeholders who wish to go further in their activities of digital inclusion. They need to adopt a new, more empathetic mindset, going beyond simple participatory design. It is thus necessary to consider the relational and social factors that influence the decision-making and behavior of citizens (Patrick and Hollenbeck 2021). However, these studies do not specifically address the design of services in the digital environment.

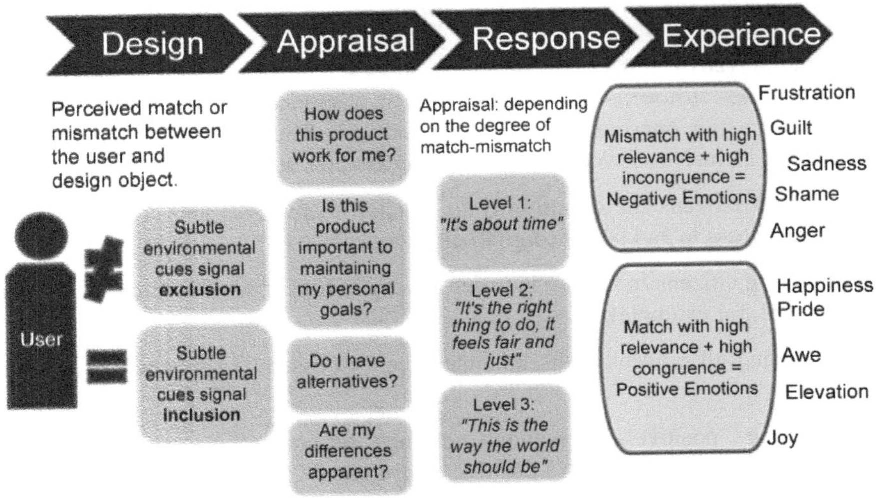

Figure 2.1. *DARE model taken from the work of Patrick and Hollenbeck (2021)*

2.4. Going further in inclusive and digital design: considering citizens' perceptions

Designing more inclusive digital experiences can enhance equal opportunities, urban attractiveness and economic efficiency. Nevertheless, the design of this type of service remains difficult in a digital context (Brotcorne et al. 2019), and organizations tend to favor certain exclusionary factors over others. Digital inclusion activities (see Table 2.1) mainly focus on tangible issues of access to and training in digital technology.

These actions do not fully consider citizens' relational and cognitive factors (interests in digital technology, expectations regarding digital technology and city services, etc.), which are just as constitutive of their feelings of inclusion or exclusion. Some research also suggests that perceived social inclusion has more of an impact than real and effective social inclusion on individual behavior (Hartung et al. 2015).

Accounting for citizen perceptions is all the more important in a digital environment as their relationship with digital technology remains ambivalent and a source of many paradoxes (Banker and Khetani 2019; Challemel Schweitzer 2019; Del et al. 2020). Research has thus emerged to specify the feeling of inclusion (Licsandru and Cui 2018; Metz and Stamper 2022), and more particularly the perception of digital inclusion (see Box 2.2).

Actions	Objectives
The Digital in Common events every year bring together a multitude of professionals, actors and communities, in order to exchange and construct strategic axes and actions in favor of the digital inclusion of tomorrow[1].	Building a virtuous and inclusive digital ecosystem.
Twenty local authorities were selected following the call for expressions of interest "Supporting local authorities to enact digital inclusion solutions for very small businesses", launched by the Banque des Territoires to strengthen the France Relance plan. The objective is to offer methodological and financial support to structure and implement the development of digital uses for VSEs in their territory.	Building a virtuous and inclusive digital ecosystem at the city level.
Since 2021 and for a period of 2 years, the state has been financing the training and deployment of 4,000 France Services digital advisors, in order to train them and enable them to lead digital initiation workshops for the 13 million French who are estranged from digital technology.	Facilitating access to digital support services.
In 2019, the state rolled out digital passes (checkbooks to be used with qualified structures to provide digital support services). Between 10,000 and 20,000 people have reportedly benefited from this system since 2019.	Facilitating access to digital support services.
The Aidants Connect system aims to secure the services of mediators who are required to take steps on behalf of the users concerned.	Securing and protecting digital support.
The France Services program was set up in response to the dematerialization of public services: the Banque des Territoires is operationalizing the France Services structures to enable support for citizens navigating administrative procedures, as close as possible to home.	Facilitating access to digital support services. Facilitating access to essential public services.
Since 2014, the Orange Foundation has offered numerous digital mediation activities alongside others for young people and women[2].	Facilitating access to digital tools. Developing digital competences.
Between December 2020 and March 2021, operation Don de Giga, launched by Bouygues Telecom, in partnership with the French Red Cross, collected more than 240,000 gigabytes, distributed nearly 20,000 packages and 10,000 smartphones to populations in need.	Facilitating access to digital tools.
SFR works in collaboration with Emmaüs Connect and WeTechCare on accessibility and digital autonomy for the population[3].	Facilitating access to digital tools.

1 See: https://numerique-en-communs.fr/.

2 See: https://www.fondationorange.com/.

Under the France very high speed plan, the public initiative network consists of setting up of very high speed public Internet networks (optical fiber) for individuals and businesses upon request from local authorities. To date, there are 463 declared RIP projects.	Bringing digital technology to all territories. Facilitating access to digital tools.
Since July 2019 and within the framework of the interministerial project "New Places, New Links", local authorities can support the development of one or more third places via calls for expressions of interest for "territorial factories". Three hundred territorial factories are planned by 2022.	Bringing digital technology to all territories.
PIX certification was developed in 2016. It supports public and private bodies in training their employees, pupils and students, for example.	Assessing, developing and certifying digital skills.

Table 2.1. *Some examples of actions promoting digital inclusion implemented in France*

Conceptualizing digital inclusion as perceived by citizens

"To make real progress on matters of diversity and inclusion [...] we must consider the human experience".

As part of a doctoral dissertation (Cauchard 2023), an exploratory qualitative study was conducted among professionals in the digital inclusion value chain and other individuals (25 interviews) in order to analyze individuals' representations and perceptions in a digital context. This study combined with a rigorous literature review proposes a conceptualization of digital inclusion as perceived by the individual in four dimensions (accessibility, external equity, autonomy and recognition).

Digital inclusion as perceived by the individual thus refers to the judgment that the user makes of: their ease of access to digital devices (accessibility); the fairness of how the organization treats the user compared to others (external fairness); the user's capacity to make use of digital technology in a free, independent and controlled manner (autonomy); the value that others give to the user's actions and digital uses (recognition). This self-assessment allows individuals to position themselves socially in relation to others and extract the value from digital resources that allows them to meet their expectations. In the long term, digital inclusion as perceived by the user promotes a better quality of life and greater well-being.

Box 2.2. *Definition of digital inclusion as perceived by the individual*

3 See: https://www.sfr.fr/options/option-solidaire.

These studies thus invite urban stakeholders to combine a collective and individual approach to digital inclusion to minimize the gap(s) between inclusive digital services and citizens' expectations.

Digital inclusion	Objectives	Dimensions
City service-level approach	Offering conditions for living together and digital pathways that promote the capacities for the participation of all citizens.	Access. Autonomy and choice. Equity. Seamless experience.
Citizen-level approach	Citizens make judgments on the conditions of living together in which they develop. These must promote their feeling of belonging to society.	Accessibility. External equity. Autonomy. Recognition.

Table 2.2. *Double approach to digital inclusion in the context of the smart city*

2.5. Conclusion: a new approach to digital inclusion for supporting citizens during the digital transition of the city

Faced with digital paradoxes and digital inequalities, smart city stakeholders must adopt a smarter approach that combines digital deployment and inclusive logic. To ensure their missions of general interest and to include citizens who have been previously excluded from digital technology, it is necessary to rethink current strategies for a more accessible digital environment. Digital inclusion is generally understood through the phenomenon of the digital divide, which emphasizes objective inequalities between citizens (conditions of access, digital skills and frequency of digital uses). In other words, digital exclusion is mainly the expression of a population's lack of education about tools and the current digital society (devices, platforms, data, etc.). The difficulties faced by citizens are thus underestimated insofar as they are assumed to concern only a marginal segment of the population (persons with disabilities, seniors, those living precariously, etc.), whereas in reality, digital equality remains difficult to achieve for the entire population despite the actions already in place: "The audiences affected are much more diverse than we might imagine [...] At Emmaüs Connect, we mainly support people between 31 and 54 years old [...]".

Given that digital inclusion strategies are meant to bring citizens closer to digital opportunities, it is essential not to limit digital inclusion to citizens' skills (digital illiteracy). Perceptions play a major role in users' behavioral responses and influence user experiences (Patrick and Hollenbeck 2021). Consequently, urban stakeholders must not only work on a more inclusive design for their digital offerings, but also incorporate digital inclusion efforts into broader social policies. This requires a better understanding of the psychological and social mechanisms involved when citizens are faced with digital pathways (social comparisons, social norms, social identity, etc.).

2.6. References

Aslam, A., Naveed, A., Shabbir, G. (2020). Is it an institution, digital or social inclusion that matters for inclusive growth? A panel data analysis. *Quality & Quantity*, 55, 333–355 [Online]. Available at: http://link.springer.com/10.1007/s11135-020-01008-3 [Accessed 15 September 2020].

Banker, S. and Khetani, S. (2019). Algorithm overdependence: How the use of algorithmic recommendation systems can increase risks to consumer well-being. *Journal of Public Policy & Marketing*, 38(4), 500–515.

Beauchamps, M. (2009). L'accessibilité numérique. *Les Cahiers du Numérique*, 5(1), 101–118.

Brotcorne, P., Bonnetier, C., Vendramin, P. (2019). Une numérisation des services d'intérêt général qui peine à inclure et à émanciper tous les usagers. *Technologie de l'information, culture & société*, 125–126.

Cauchard, L. (2023). L'inclusion numérique perçue : conceptualisation, antécédents et conséquences sur le parcours utilisateur et l'engagement dans la relation. PhD Thesis, Université de Montpellier, Montpellier [Online]. Available at: http://www.theses.fr/s228294 [Accessed 14 June 2023].

Challemel Schweitzer, V. (2019). Paradoxes de l'empowerment du consommateur participatif : le cas des bornes digitales d'accès au service. In *35e Congrès de l'Association Française du Marketing*, Le Havre.

Cohen, E.B. (2012). *Issues in Informing Science & Information Technology*, volume 9. Informing Science, Santa Rosa.

Cushman, M. and Klecun, E. (2006). How (can) nonusers engage with technology: Bringing in the digitally excluded. In *Social Inclusion: Societal and Organizational Implications for Information Systems*, Trauth, E.M., Trauth, E.M., Howcroft, D., Butler, T., Fitzgerald, B., DeGross, J.I. (eds). Springer, Boston.

Dagenais, D. (2012). Littératies multimodales et perspectives critiques. Recherches en didactique des langues et des cultures. *Les Cahiers de l'Acedle*, 9(2) [Online]. Available at: https://journals.openedition.org/rdlc/2338 [Accessed 15 November 2021].

Del Bucchia, C., Lancelot Miltgen, C., Russell, C.A., Burlat, C. (2020). Empowerment as latent vulnerability in techno-mediated consumption journeys. *Journal of Business Research*, 124, 629–651 [Online]. Available at: http://www.sciencedirect.com/science/article/pii/S0148296320301788 [Accessed 16 April 2020].

Dennis, C., Alamanos, E., Papagiannidis, S., Bourlakis, M. (2016). Does social exclusion influence multiple channel use? The interconnections with community, happiness, and well-being. *Journal of Business Research*, 69(3), 1061–1070.

Durand-Mégret, B., Ezan, P., Vanheems, R. (2016). Vers une nouvelle manière de concevoir la compétence et le risque perçu de l'acheteur à l'ère du numérique. Le cas de l'adolescent au sein de la famille. *Management & avenir*, 83(1), 33.

Fisk, R.P., Dean, A.M., Alkire, L., Joubert, A., Previte, J., Robertson, N., Rosenbaum, M.S. (2018). Design for service inclusion: Creating inclusive service systems by 2050. *Journal of Service Management*, 29(5), 834–858.

Gardou, C. (2012). *La société inclusive, parlons-en !* Éditions Érès, Toulouse [Online]. Available at: https://www.editions-eres.com/ouvrage/3059/la-societe-inclusive-parlons-en [Accessed 15 November 2021].

Ghorra-Gobin, C. (2018). Smart city. *Quaderni*, 96(2), 5–15.

Granjon, F. (2009). Inégalités numériques et reconnaissance sociale. *Les Cahiers du Numérique*, 5(1), 19–44.

Granjon, F. (2010). Le "non-usage" de l'internet : reconnaissance, mépris et idéologie. *Questions de communication*, 2010/2(18), 37–62.

Haferkamp, N. and Krämer, N.C. (2011). Social comparison 2.0: Examining the effects of online profiles on social-networking sites. *Cyberpsychology, Behavior, and Social Networking*, 14(5), 309–314.

Hargittai, E. (2002). Second-level digital divide: Differences in people's online skills. *First Monday*, 7(4) [Online]. Available at: https://firstmonday.org/ojs/index.php/fm/article/view/942 [Accessed 26 November 2019].

Hargittai, E. and Hinnant, A. (2008). Digital inequality: Differences in young adults' use of the internet. *Communication Research*, 35(5), 602–621.

Hartung, F.-M., Sproesser, G., Renner, B. (2015). Being and feeling liked by others: How social inclusion impacts health. *Psychology & Health*, 30(9), 1103–1115.

Jauréguiberry, F. (2003). *Internet comme espace inédit de construction de soi*. L'Harmattan, Paris.

Jauréguiberry, F. (2005). *L'immédiateté télécommunicationnelle*. Éditions de l'Aube, La Tour d'Aigues.

Jauréguiberry, F. (2012). Retour sur les théories du non-usage des technologies de communication. In *Connexions : communication numérique et lien social*, Klein, S.P. and Klein, A. (eds). Presses universitaires de Namur, Namur.

Katz, J.E. and Aspden, P. (1998). Internet dropouts in the USA: The invisible group. *Telecommunications Policy*, 22(4), 327–339.

Kellner, C., Massou, L., Morelli, P. (2010). (Re)penser le non-usage des tic. *Questions de communication*, 18(2), 7–20.

Kenyon, S., Lyons, G., Rafferty, J. (2002). Transport and social exclusion: Investigating the possibility of promoting inclusion through virtual mobility. *Journal of Transport Geography*, 10(3), 207–219.

Kiyindou, A. (2007). De la fracture numérique à la fracture cognitive : pour une nouvelle approche de la société de l'information. In *Thémat'IC 2007 "La maîtrise de l'information par les adultes : enjeux et méthodes"*, Strasbourg [Online]. Available at: http://www.enssib.fr/bibliotheque-numerique/notice-1948.

Licsandru, T.C. and Cui, C.C. (2018). Subjective social inclusion: A conceptual critique for socially inclusive marketing. *Journal of Business Research*, 82, 330–339.

Livingstone, S. and Helsper, E. (2007). Gradations in digital inclusion: Children, young people and the digital divide. *New Media & Society*, 9(4), 671–696.

Mazet, P. (2019). Vers l'État plateforme. La vie des idées, 2 April [Online]. Available at: https://laviedesidees.fr/Vers-l-Etat-plateforme.

Metz, I., Stamper, C.L., Ng, E. (2022). Feeling included and excluded in organizations: The role of human and social capital. *Journal of Business Research*, 142, 122–137.

Meuter, M., Ostrom, A., Roundtree, R., Bitner, M. (2000). Self-service technologies: Understanding customer satisfaction with technology-based service encounters. *Journal of Marketing*, 64(3), 50–64.

Patrick, V.M. and Hollenbeck, C.R. (2021). Designing for all: Consumer response to inclusive design. *Journal of Consumer Psychology*, 31(2), 360–381.

Peter, J. and Valkenburg, P.M. (2006). Adolescents' internet use: Testing the "disappearing digital divide" versus the "emerging digital differentiation" approach. *Poetics*, 34(4), 293–305.

Pinède, N. (2018). Numérique et situations de handicap : les enjeux de l'accessibilité. Introduction. *tic&société*, 12(2), 1–8.

Vodoz, L. (2010). Fracture numérique, fracture sociale : aux frontières de l'intégration et de l'exclusion. *Sociologies* [Online]. Available at: http://journals.openedition.org/sociologies/3333 [Accessed 22 January 2020].

Wallcook, S. and Morris, K. (2017). Supporting digital health and digital inclusion: An occupational therapy perspective. *Journal of Corporate Citizenship*, 68, 82–94.

Wilson, C. and Secker, J. (2015). Validation of the social inclusion scale with students. *Social Inclusion*, 3(4), 52–62.

Wyatt, S. (2003). Non-users also matter: The construction of users and non-users of the Internet. In *How Users Matter: The Co-Construction of Users and Technology*, Oudshoorn, N. and Pinch, T. (eds). MIT Press, Cambridge.

Yu, B., Ndumu, A., Mon, L.M., Fan, Z. (2018). E-inclusion or digital divide: An integrated model of digital inequality. *Journal of Documentation*, 74(3), 552–574.

3

Co-creating

3.1. Introduction: the smart city, toward a new model of governance

According to Nesti, while there is no academic consensus on the definition of the smart city, it cannot be defined exclusively by its technological component, which reflects the transition "from an IT city to a digital city" (Nesti 2020). Indeed, recent literature shows a broader acceptance of the "holistic" model of the smart city, which privileges quality of life and sustainability over a model focused on technology (Sharifi 2019). We therefore adopt Bizien's (2020) definition: "Smart cities involve a connection between urban infrastructure and their citizens, creating information which is used through ICT to improve the overall quality of life of residents".

Meijer and Rodriguez-Bolivar identify three axes of research on the smart city: it uses smart technologies (technological angle), employs smart personnel (human resources angle) and promotes smart collaborations (governance angle) (Meijer and Rodriguez-Bolivar 2016). We are adopting this third approach, focusing on interactions or even collaboration between stakeholders in the city that integrates citizen expectations. We will focus more particularly on its new modes of governance. Combining the contributions of theories of public value and participatory democracy, we propose a framework for analyzing these modes of governance that sheds new light on value creation (Moore 1995) through citizen participation in this governance. A literature review will lead us to identify the key characteristics of public value creation and participatory democracy and to suggest an analytical framework that matches each set of characteristics of the different smart cities studied.

Chapter written by Isabelle BORIES-AZEAU and Jessica CLEMENT.

3.2. A proposed framework of analysis

3.2.1. *Urban governance: networked governance*

Governance is a polysemic concept which covers all interactions between the various public and private actors in developing and steering public policies (Le Galès 2006). Urban governance refers to a collaborative process between governmental and nongovernmental stakeholders in developing urban public policies (Blanco 2014). For the smart city, this means a multi-stakeholder and networked urban collaboration that can create local public value (Barrutia et al. 2022). In this context, the public stakeholder is positioned as facilitator and regulator of an ecosystem of stakeholders (Moore 1993) creating links between the territory's resources. The objective is to construct an environment that encourages the expression of each stakeholder within a collective project. The governance of smart cities is thus characterized by horizontality and a multi-stakeholder approach.

Smart cities have also been presented as collaborative ecosystems that improve the quality of residential life (Appio et al. 2019) and facilitate innovation through links between stakeholders. For Barrutia et al. (2022), practices of urban innovation toward a smart city generate public value. But as local governments cannot create this value alone, new collaborative modes of governance must be established (Barrutia et al. 2022) with the participation of citizens (De Guimarães et al. 2020) and stakeholders (Freeman 1984).

The question of leadership, particularly political leadership, may also arise. Political leadership covers the intervention of elected officials in the production and steering of public policies and their capacity to act, that is, to coordinate actions (Sawicki 2003), with a strong political engagement. The political leader satisfies the interests of the main stakeholders (Nutt and Backoff 1993). Nutt and Backoff (1993) emphasize the specific context and collective dimension of political leadership. Brookes (2008) insists on shared, collaborative leadership between different organizations with a common vision and goals, which would favor greater effectiveness (Pearce 1999). Morse (2010, p. 235) joins this approach with the concept of integrative public leadership: "[…] the process through which multiple actors with multiple interests create a common vision and work together to create public value (the common good or public interest)".

3.2.2. *Creation of public value, a challenge for participatory democracy*

For Moore (1995), public value covers all of the societal benefits produced by the activity of managers in the public service. The conceptual framework of the strategic triangle (Moore 1995) allows us to understand strategic challenges and the corresponding choices through three questions:

1) What public value does the organization seek to produce (strategic objectives of the organization)?

2) What sources of legitimacy and support are needed to enable the organization to take measures to create this value (legal framework of action and political legitimation)?

3) What operational capacities, such as investments, innovations or alliances, are needed to achieve the desired results?

The creation of public value requires interactions between these three poles, but also the public support and legitimacy provided by political leadership.

The strategic triangle aims to provide public managers with a basis for diagnosis of the territory to identify opportunities for value creation and propose an action plan to exploit them (Moore 2021). The objective is then to mobilize the dynamic capacities of the organization (Teece 2012) to adapt to a constantly changing environment.

The coexistence of political leadership with managerial leadership is also to be encouraged, integrating the roles of political and administrative leaders into urban management (Carassus and Gardey 2009). If politicians have a territorial vision of the management of public policies, the administrative authorities have an organizational, managerial and instrumental understanding aimed at the rational use of local resources.

A new model of governance for smart cities is therefore emerging that accounts for public value and participatory democracy and their perception by public managers. Participatory democracy refers to "a process of collective decision-making combining elements of direct democracy and representative democracy. Citizens have the power to decide on policy proposals and politicians assume the role of policy implementors" (Aragonès and Sánchez-Pagés 2009, p. 56). It supports a system of governance in which citizens can influence the development of public policy (Shadowen et al. 2020). It allows for the learning of democratic structures (Pateman 2012) and therefore appears as a social value in itself (Dachler and Wilpert 1978), especially when citizens deliberate directly within local government structures on issues that concern them (Fuchs 2007). The aim is to promote horizontal public action, broadly involving stakeholders, particularly users, in governance (Richez-Battesti and Valade 2017). Citizenship through proximity, that is to say the direct participation of users in the creation of public services, thus recognizes their needs and identifies unrecognized needs.

	Public value	Participatory democracy
Evaluation of governance	(1) Clear definition of the "public value" that the local government seeks to produce.	(7) Continuing role in participation.
	(2) Achievement of results (a).	(8) Right of all residents to participate in decision-making.
	(3) The public value created is pursued in the context of residents' needs.	(9) Participation is seen as a means to an end, as it enhances empowerment and capacity.
	(4) Proposal or affirmation of legitimacy and support, increasing confidence.	(10) Participation at local level.
	(5) Identification and realization of (or plan to realize) organizational capacities.	(11) Residents have the opportunity to have an impact on public policy at the time of its development.
	(6) Public value created with maximum benefit using minimum resources (efficiency).	(12) Clear leadership to guide processes of participation.

Table 3.1. *Analytical framework for the key components of governance of smart cities. (a) The extent to which a public body improves publicly valued outcomes in a wide variety of areas*

The transition to the smart city adopts technological tools that improve the quality of life of residents, thus changing the organization of municipal activities and the provision of services to adapt to developments in their context (Paulin 2018). Cities and territories become smarter when they use technology appropriately, but also when they monitor, understand and plan better using new tools (Meijer and Bolívar 2016). This implies digital levers for redeployment of resources (e-services and e-administration) and a repositioning of the partnership strategy, notably including the integration of changes in the associative fabric.

Through information and communication technologies, dematerialization has favored e-participation or "the voluntary engagement of citizens and their participation in public administration affairs and public decision-making through the use of government-provided web applications" (Kim and Lee 2012, p. 820).

From this literature review, we have identified the key components of public value creation and participatory democracy that can be mobilized to evaluate smart city governance. Table 3.1 presents this analytical framework.

3.3. Methodology

The methodology we have mobilized has previously been used in studies seeking to demonstrate how smart city approaches correspond to certain criteria of governance (Clement and Crutzen 2021). Its steps are detailed in Figure 3.1.

We first identified the cases to be studied based on several criteria: appearances in the international rankings of smart cities (IMD 2021) and in academic articles and chapters that suggest that these cities are distinguished by "best practices". We also conducted a study of documentation on the smart city initiative (policy documents, reports, etc.) to enrich our research material. We were thus able to select six cities: Amsterdam (Netherlands), Johannesburg (South Africa), London (England), Melbourne (Australia), Montpellier (France) and Ottawa (Canada).

Second, we searched for documents between July and August 2022 for each case: academic articles or chapters, reports, government policy or strategy documents, press articles and websites. The documents were chosen based on keyword searches related to "smart city", "public value", "participation", "participatory democracy", and "governance". We therefore cover both the elements of the smart city concept in a city and the way in which it is conceptualized and implemented, as well as related policies. The documents were then screened for relevance. Relevant documents for evaluating governance are summarized in Box 3.1, and Table 3.2 shows the number and type of document for each case. A total of 247 documents were collected. Although every document was used to inform the framework, only the most exemplary ones are cited in the case studies.

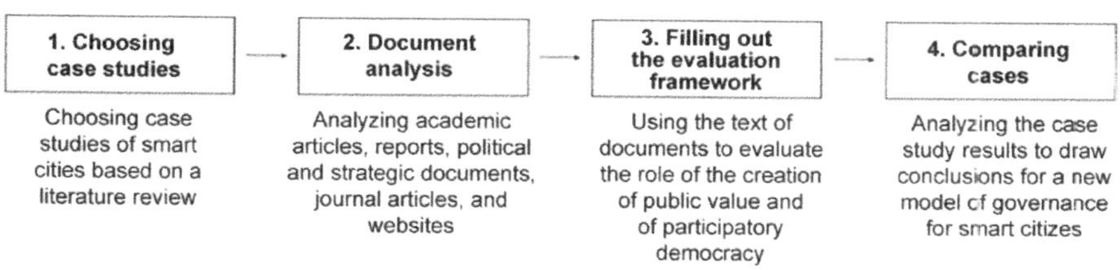

Figure 3.1. *Research strategy and methodology*

Amsterdam Smart City (n.d.). Citizens & living. Amsterdam Economic Board [Online]. Available at: https://amsterdamsmartcity.com/channel/citizens-living.

Auci, S. and Mundula, L. (2012). Smart cities and a stochastic frontier analysis: A comparison among European cities. Report, SSRN.

City for All Women Initiative (2012). Making Ottawa's municipal governance inclusive. Report, CAWI, Ottawa-Gatineau.

City of Johannesburg (2013a). State of the city address by the executive mayor of the City of Johannesburg. Councilor Mpho Parks Tau, Linder Auditorium. Speech, University of Witwatersrand, Johannesburg.

City of Johannesburg (2013b). Summarized version of 2013/2016 Integrated Development Plan (IDP). Report, City of Johannesburg, Johannesburg.

City of Johannesburg (2014). 2014 State of the city address by the executive mayor of Johannesburg Clr Mpho Parks Tau. Report, City of Johannesburg, Johannesburg.

City of Johannesburg (2015). State of the city address 2015, delivered by the executive mayor, Councilor Mpho Parks Tau. Report, City of Johannesburg, Johannesburg.

City of Johannesburg (2016). Tabling of the 2016/17 institutional Services Delivery Budget and Implementation Plan (SDBIP). Report, City of Johannesburg, Johannesburg.

City of Melbourne (n.d.). Projects | Participate Melbourne. Report, City of Melbourne, Melbourne [Online]. Available at: https://participate.melbourne.vic.gov.au/projects.

City of Melbourne (2016). Refreshing the future Melbourne plan: Citizens' jury report. Report, City of Melbourne, Melbourne.

City of Ottawa (2017). Smart City 2.0. Report, City of Ottawa, Ottawa.

Greater London Authority (2016). DD1455 GLA signatory to HM government's digital inclusion charter. Report, Greater London Authority, London.

Greater London Authority (2018). Smarter London together. Report, Greater London Authority, London.

IMD (2021). Smart city index 2021: City performance overview. Report, IMD.

Montpellier Méditerranée Métropole (2022). [Online]. Available at: https://www.montpellier3m.fr/actualite/appel-projets-innovants-au-service-de-la-ville-intelligente.

Steen, K. and Van Bueren, E. (2017). *Urban Living Labs: A Living Lab Way of Working*. Amsterdam Institute for Advanced Metropolitan Solutions, Amsterdam.

Box 3.1. *The documents studied*

Type of document	Amsterdam		Johannesburg		London		Melbourne		Montpellier		Ottawa	
	N	%	N	%	N	%	N	%	N	%	N	%
Academic article/chapter	15	35.7	6	14.3	9	22.5	7	18.9	6	12.5	7	17.9
Report	0	0	0	0.0	0	0.0	2	5.4	0	0	0	0
Government document	6	14.3	32	76.2	29	72.5	24	64.9	34	70.8	29	74.4
Smart city platform document	14	33.3	0	0	0	0.0	0	0	0	0	0	0
Press article	5	11.9	4	9.5	2	5	2	5.4	8	16.7	1	2.6
Websites	2	4.8	0	0	0	0	2	5.4	0	0	2	5.1
Total	42	100	42	100	40	100	37	100	48	100	39	100

Table 3.2. *Number and type of documents for each city studied*

3.4. Results

Concerning public value creation, for the first component, namely a clear definition of the public value that the local government aims to produce, the criteria were met if a city clearly stated its objectives for the transition to a smart city. Our results show that five out of six cities had either formal smart city strategies or a clear statement of the public value at which local government was aiming. The second component concerned the achievement of results, that is, an improvement in results appreciated by the public in various areas. Here, we find support for all cities. This shows the transversality of the smart city. For the third component, five out of six cities clearly indicated that the creation of public value responds to the needs of citizens. Four out of six cities are characterized by the proposal or affirmation of legitimacy and support and the objective of gaining their residents' trust, the fourth component of governance. Component five considered whether a city identified the importance of its organizational capacities or planned to improve them. All six cities showed these characteristics in their governance. Finally, only three out of six cities had documents corresponding to the sixth component, that is, the most effective possible creation of public value.

For the second part of the framework concerning participatory democracy, component seven identified a continuing role in participation. All six cities had at least some measures in place to promote this. Regarding the eighth component, all six cities had established the right of everyone to participate in decision-making. For the ninth component, four cities saw participation as a means to achieve their goals, as improving empowerment and capacities. For the 10th component, local participation, all six cities qualified. Only in five out of six cities do we find the last two components: component 11, dealing with the possibility of residents having an impact on creating public policy, and component 12, clear leadership guiding processes of participation.

A table presents each case with explanations of the components identified.

3.4.1. *Amsterdam*

Amsterdam is one of the three smart city initiatives that have addressed every component in the governance framework. This is an interesting result, as Amsterdam's smart city program is managed by a multi-stakeholder platform (Amsterdam Smart City) that includes the local government, but is not controlled by it. This program was in fact developed in 2007 by three key stakeholders: the municipal government, Amsterdam Innovation Motor (AIM) and Liander, a Dutch energy company. The three stakeholders collaborated to support the phases of development of smart cities: initial planning, project development, project monitoring and evaluation and communication of results (Mora and Bolici 2017). Thus, the governance of Amsterdam is envisaged at the same time through the local government, its actions and the platform.

Evaluation of governance	
Public value	**Participatory democracy**
Clear objectives as a smart city (Mora and Bolici 2017), including reducing carbon emissions (Angelidou 2016).	Stimulation of and support for cooperation between the public and private sectors and citizen participation in each project (Mora and Bolici 2017).
The multi-stakeholder platform covers various areas: smart living, smart working, open data and sustainable property (Yigitcanlar et al. 2019).	Sharing each resident's ideas on the smart city platform (Amsterdam Smart City n.d.).
Engaging the local community in a bottom-up approach to choosing solutions; examples: pilot projects for testing solutions (Yigitcanlar et al. 2019) or Living Labs (Mora and Bolici 2017).	Raising awareness using applications: a key feature of the smart city program; example: culture change around energy consumption.
Trust, a key element of the Living Lab program (Steen and Van Bueren 2017).	Residents participate at the local level.
The platform aims to address challenges with input from relevant stakeholders to optimize capacity (Camboim et al. 2019).	Residents can impact the development of public policies in some cases (Angelidou 2016).
Projects are often tested through (small-scale) pilots before possibly being extended (Angelidou 2014).	The platform coordinates stakeholders, strategy deployment, and projects (Angelidou 2016; Mora and Bolici 2017).

Table 3.3. *Explanations relating to the components for Amsterdam*

To conclude, we note that while several sources suggest that all residents can participate in decision-making – in one way or another – documents show that practice is different (Leclercq and Rijshouwer 2022). This implies obstacles in deployment of the strategy or a guidance document.

3.4.2. *Johannesburg*

The city of Johannesburg has been working on its smart city approach since 2010, when it was defined as a goal of the city's strategic development plan. Johannesburg first adopted a smart city strategy with a roadmap in 2014. The 2019 revised version aims to transform Johannesburg into a citizen-centered and inclusive city through the use of digital technologies. The smart city is also seen as a means of promoting the territory's sustainability and resilience. Like Amsterdam, Johannesburg includes all of the components of governance and public value.

Evaluation of governance	
Public value	**Participatory democracy**
Clear definition of the public value that the local government aims to produce with the smart city strategy. Proposal of a roadmap to achieve the objectives.	All local initiatives, particularly in the case of the smart city, involve citizens in a logic of continuous engagement as new initiatives emerge.
Adoption of a program covering several areas: economic development, poverty eradication, housing, sustainability, governance, etc. (Bwalya 2019).	Organizing public meetings on various issues at neighborhood level to strengthen participatory democracy (Winkler 2011).
Choice of areas linked to residents' needs: delivering benefits (City of Johannesburg 2013a).	*Vulindlel' eJozi* project to support the participation and integration of young people into the economy and society (City of Johannesburg 2015).
Relationship of trust between government and citizens, notably with an independent mediator (City of Johannesburg 2014).	Local participation, or even "active participation in deliberation" (Peyroux 2015, p. 568) for developing local policies and strategies.
Strengthening the capacity to deploy effective and efficient ICT infrastructure (City of Johannesburg 2013b).	Organizing public consultations during the drafting of government documents.
Providing services that are easy to access and use, but also efficient, responsive and transparent (City of Johannesburg 2016).	Clear leadership, generally from local specialized departments, and from a central team for the smart city, guides the participation processes.

Table 3.4. *Explanations relating to the components for Johannesburg*

3.4.3. *London*

The first smart city strategy, the *Smart London Plan*, dates from 2013 and was updated in 2016. A new strategy, *Smarter London Together*, was published in 2018. The latest strategy is structured around five missions: better services for users, a new deal for city data, world-class connectivity and smarter streets, strengthening digital leadership and skills, and improving city-wide collaboration. The accent is therefore on collaboration and participation to encourage uptake of digital technologies by residents.

Evaluation of governance	
Public value	**Participatory democracy**
The smart city program is based on a series of strategies between 2013 and 2018 with clear objectives (Greater London Authority 2018) and varying themes.	Continuous solicitation of participation through the Talk London platform with an ongoing debate feeding into policy on a range of relevant topics.
Online monitoring of progress on these themes.	Platform open to all residents.
The strategy and related initiatives were developed after a "listening tour", so that the public value created would emerge from the needs of residents.	Digital inclusion plan to develop online services and improve overall quality of life in a new digital future (Greater London Authority 2016).
The city is trying, through a search for data trusts, to encourage data sharing between data-holding and data-using organizations (Greater London Authority 2018).	Different types of initiatives illustrate local participation.
A digital leadership program ensures that leaders have a basic understanding of services involving technology and its social impact.	Citizen contribution (Crowdfund London digital platform): submitting and developing an idea, coordinating local support, resources, and funding via a public campaign (Greater London Authority 2018).
Collaboration with local stakeholders. Engaging citizens in solving urban problems and developing more efficient services (Bibri and Krogstie 2020).	The smart city program is a directive from the mayor, with clear leadership to guide related initiatives and the participatory processes that underpin them.

Table 3.5. *Explanations relating to the components for London*

3.4.4. *Melbourne*

Melbourne's smart city approach was initiated as early as 2004 when the Melbourne City Council Plan 2004–2008 included a strategic objective aiming to make Melbourne a smart city. This is currently linked to the *Knowledge City* initiative (Leorke and Wyatt 2018), focusing on the collective power of citizens to improve the city's quality of life.

Evaluation of governance	
Public value	**Participatory democracy**
Lack of an official strategy (Clement and Crutzen 2021), and thus of a clear definition of public value.	Continuous participation through a design thinking process (Anthopoulos and Reddick 2016).
But the smart city program covers various domains (Irajifar 2022).	Open platform of the Participate Melbourne program: resident participation in decision-making, at least in this context (City of Melbourne n.d.).
Several initiatives aimed at responding to citizens' needs and improving their engagement (Irajifar 2022).	Local participation.
–	Citizen juries: residents can influence public policymaking (Vein 2017).
Example of the *Digital Twin* initiative (intelligent data management): predictive capacity to respond to disasters or shocks.	Direct contribution of residents to creation of local policies in citizens' juries (Vein 2017).
–	Clear leadership from the Mayor structured by the Future Melbourne 2026 plan (City of Melbourne 2016).

Table 3.6. *Explanations relating to the components for Melbourne*

3.4.5. *Montpellier (Montpellier metropolitan area)*

Montpellier has long had a reputation in France as a city of knowledge, as it is home to several universities and large international companies, leading to the creation of "poles of excellence" (Bakis and Schon 2012). This dynamic has supported the smart city approach.

The city is now often cited as a reference for European smart cities (Auci and Mundula 2012). The smart city of Montpellier is envisaged based on a transversal approach focused on supporting residents' quality of life, the economy and sustainability.

Evaluation of governance	
Public value	**Participatory democracy**
Clear definition of public value with three key objectives for the smart city.	Continuous participation of citizens involved throughout the whole process.
Transversal approach to the smart city.	Citizens encouraged to express themselves directly on key issues, especially on the development of the Smart City program.
Creation of public value to respond to residents' needs, particularly in key areas such as health and mobility.	–
–	Active local participation.
Developing organizational capacities by improving internal governance in certain practices, such as purchasing policies.	Support for participation: organizing an open workshop on the design of public policies, participation of citizens in developing new public services and feedback on the management of public innovation.
–	Smart City program, led by the Montpellier metropolitan area, with calls for projects and initiatives linked to participation.

Table 3.7. *Explanations relating to the components for Montpellier*

3.4.6. *Ottawa*

Ottawa's smart city approach is supported by its "Smart City 2.0" strategy of 2017. This was developed after an official stakeholder workshop in 2016 which gave rise to an initial set of specifications. The strategy has three objectives: developing a connected city, a smart economy and an innovative government.

Interestingly, in 2003, Ottawa developed a "public participation policy" after community consultations with residents. However, a 2012 report suggests that it had not been fully implemented and no other documents with these keywords were found in academic articles or on the City of Ottawa website after 2012 (City for All Women Initiative 2012).

Evaluation of governance	
Public value	**Participatory democracy**
Strategy underpinned by a clear definition of the public value that the local government wants to produce.	Several initiatives included in the strategy to ensure continued engagement: Hackathons, online engagement tools and Meet Ups.
Several areas covered: energy, communication, transport and education. The strategy is linked to other local strategies (City of Ottawa 2017).	Cases where citizens were able to comment on draft programs and strategies before their final version.
–	–
A program for open government to improve opportunities and transparency between the city and its residents: an open data site, responses to residents' requests for data and information.	Local participation.
The strategy mobilizes a multi-stakeholder working group to provide capacity, which involves improving organizational capacities.	–
–	The city manages a website dedicated to engagement opportunities, Engage Ottawa.

Table 3.8. *Explanations relating to the components for Ottawa*

3.5. Conclusion

This chapter addresses the concepts of public value and participatory democracy, distinguished in the literature as two dimensions of the governance of smart cities. Our analytical framework suggests combining these concepts by postulating that participatory democracy promotes the creation of public value. We then identified several cases of "best practices" of smart cities in the literature and international rankings. Then, using an in-depth analysis of academic literature and local government documentation, we evaluated how each city integrated the governance components of our framework.

Our results suggest that some components of public value and participatory democracy are broadly present in current approaches to smart cities. That said, governance in smart cities tends to focus more on participatory democracy. This is an interesting result, as recent literature has underlined the importance of citizen participation in smart city initiatives (Meijer and Bolívar 2016; De Guimarães et al. 2020). The components related to public value are less emphasized, particularly regarding the question of its efficient production, but also that of gaining the confidence of residents. Overall, our results suggest that some cases of smart city best practices tend to adopt all or several of the components of governance. There remains, however, obstacles, such as improving access to the digital technologies upon which participation often relies.

Our study presents limitations. First, despite the exhaustiveness of our literature review for building the governance analysis framework, we consider it to be an adaptive framework that could be complemented by other components that integrate the new governance challenges of smart cities. Thus, the issue of administrative fragmentation as an obstacle to adoption of smart city initiatives (Kitchin and Moore-Cherry 2021) could be studied in more depth. Second, we evaluated six cities in different parts of the world. However, in order to generalize this framework, more contexts could be studied using the assessment framework. We therefore invite future studies to consider new contexts. However, the framework offers a new way to evaluate smart city governance and to ensure that cities aim to create public value and include citizen participation in their smart city approach.

3.6. References

Angelidou, M. (2014). Smart city policies: A spatial approach. *Cities*, 41(1), S3–S11.

Angelidou, M. (2016). Four European smart city strategies. *International Journal of Social Science Studies*, 4(4), 18–30.

Appio, F., Lima, M., Paroutis, S. (2019). Understanding smart cities: Innovation eco-systems, technological advancements, and societal challenges. *Technological Forecasting and Social Change*, 142, 1–14.

Aragonès, E. and Sánchez-Pagés, S. (2009). A theory of participatory democracy based on the real case of Porto Alegre. *European Economic Review*, 53(1), 56–72.

Bachrach, P. and Botwinick, A. (1992). Power and empowerment: A radical theory of participatory democracy. PhD Thesis, Temple University, Temple.

Bakis, H. and Schon, A. (2012). Ville de la connaissance et terreau numérique. *NETCOM*, 26(3/4), 275–306.

Barrutia, J., Echebarria, C., Aguado-Moralejo, I., Apaolaza-Ibáñez, V., Hartmann, P. (2022). Leading smart city projects: Government dynamic capabilities and public value creation. *Technological Forecasting and Social Change*, 179, 121679.

Bibri, S. and Krogstie, J. (2020). The emerging data–driven smart city and its innovative applied solutions for sustainability: The cases of London and Barcelona. *Energy Informatics*, 3(1), 1–42.

Bizien, C. (2020). Evolutions réglementaires et logique d'acteur : l'institutionnalisation des villes intelligentes. PhD Thesis, Université Paris 1, Paris.

Blanco, I. (2014). Between democratic network governance and neoliberalism: A regime-theoretical analysis of collaboration in Barcelona. *Cities*, 44, 123–130.

Brookes, S. (2008). Are public leaders up to standard? *CSL Leadership Review*, 1(4), 211–224.

Bwalya, K. (2019). The smart city of Johannesburg, South Africa. In *Smart City Emergence: Cases from Around the World*, Anthopoulos, L. (ed.). Elsevier, Amsterdam.

Camboim, G., Zawislak, P., Pufal, N. (2019). Driving elements to make cities smarter: Evidences from European projects. *Technological Forecasting and Social Change*, 142, 154–167.

Carassus, D. and Gardey, D. (2009). Une analyse de la gestion de la performance par les collectivités locales françaises : un modèle administratif ou politique ? *Revue française de finances publiques*, 107, 101–129.

Clement, J. and Crutzen, N. (2021). How local policy priorities set the smart city agenda. *Technological Forecasting and Social Change*, 171, 120985.

Dachler, P. and Wilpert, B. (1978). Conceptual dimensions and boundaries of participation in organizations: A critical evaluation. *Administrative Science Quarterly*, 23(1), 1–39.

De Guimarães, J., Severo, E., Felix Júnior, L., Da Costa, W., Salmoria, F. (2020). Governance and quality of life in smart cities: Towards sustainable development goals. *Journal of Cleaner Production*, 253, 119926.

Freeman, R. (1984). *Strategic Management: A Stakeholder Approach*. Cambridge University Press, Cambridge.

Fuchs, D. (2007). Participatory, liberal and electronic democracy. In *Participatory Democracy and Political Participation: Can Participatory Engineering Bring Citizens Back In?*, Zittel, T. and Fuchs, D. (eds). Routledge, London.

Irajifar, L. (2022). Digital solutions for resilient cities: A critical assessment of resilience in smart city initiatives in Melbourne, Victoria. In *Resilient Smart Cities*, Sharifi, A. and Salehi, P. (eds). Springer, Cham.

Katsonis, M. (2019). Designing effective public engagement: The case study of future Melbourne 2026. *Policy Design and Practice*, 2(2), 215–228.

Kim, S. and Lee, J. (2012). E-participation, transparency, and trust in local government. *Public Administration Review*, 72(6), 819–828.

Kitchin, R. and Moore-Cherry, N. (2021). Fragmented governance, the urban data ecosystem and smart city-regions: The case of Metropolitan Boston. *Regional Studies*, 55(12), 1913–1923.

Le Galès, P. (2006). *Gouvernement et gouvernance des territoires. Problèmes politiques et sociaux*. La Documentation française, Paris.

Leclercq, E. and Rijshouwer, E. (2022). Enabling citizens' right to the smart city through the co-creation of digital platforms urban transformations. *Urban Transformations*, 4(2), 1–19.

Leorke, D. and Wyatt, D. (2018). *Public Libraries in the Smart City*. Palgrave Macmillan, London.

Meijer, A. and Bolívar, M. (2016). Governing the smart city: A review of the literature on smart urban governance. *International Review of Administrative Sciences*, 82(2), 392–408.

Moore, J. (1993). Predators and prey: A new ecology of competition. *Harvard Business Review*, 71(3), 75–86.

Moore, M. (1995). *Creating Public Value: Strategic Management in Government*. Harvard University Press, Cambridge.

Moore, M. (2021). Creating public value: The core idea of strategic management in government. *International Journal of Professional Business Review*, 6(1), 1–2.

Mora, L. and Bolici, R. (2017). How to become a smart city: Learning from Amsterdam. In *Smart and Sustainable Planning for Cities and Regions*, Bisello, A., Vettorato, D., Stephens, R., Elisei, P. (eds). Springer, New York.

Morse, R. (2010). Integrative public leadership: Catalyzing collaboration to create public value. *The Leadership Quarterly*, 21(2), 231–245.

Nesti, G. (2020). Définir et évaluer la nature transformationnelle de la gouvernance des villes intelligentes : observations issues de quatre cas européens. *Revue internationale des sciences administratives*, 86(1), 23.

Nutt, P. and Backoff, R. (1992). *Strategic Management of Public and Third Sector Organizations*. Josey-Bass Publishers, Hoboken.

Pateman, C. (2012). Participatory democracy revisited. *Perspectives on Politics*, 10(1), 7–19.

Paulin, A. (2018). *Smart City Governance*. Elsevier, Amsterdam.

Pearce, C. (1999). The relative influence of vertical vs. shared leadership on the longitudinal effectiveness of change management teams. In *Conference at the Academy of Management*, Chicago.

Peyroux, E. (2015). Discourse of urban resilience and "inclusive development" in the Johannesburg growth and development strategy 2040. *European Journal of Development Research*, 27(4), 560–573.

Richez-Battesti, N. and Vallade, D. (2017). ESS, territoire et régime de solidarité : l'innovation sociale comme levier pour une "ville solidaire" ? *Management & avenir*, 97(7), 105–127.

Sawicki, F. (2003). Leadership politique : un concept à remettre sur le métier ? In *Le leadership politique et les territoires. Les cadres d'analyse en débat*, Smith, A. and Sorbets, C. (eds). Presses Universitaires de Rennes, Rennes.

Shadowen, N., Lodato, T., Loi, D. (2020). Participatory governance in smart cities: Future scenarios and opportunities. In *Distributed, Ambient and Pervasive Interactions (HCII 2020)*. Springer, Cham.

Sharifi, A. (2019). A critical review of selected smart city assessment tools and indicator sets. *Journal of Cleaner Production*, 233, 1269–1283.

Teece, D. (2012). Dynamic capabilities: Routines versus entrepreneurial action. *Journal of Management Studies*, 49(8), 1395–1401.

Vein, C. (2017). Melbourne, Barcelona, San Francisco: Three smart cities changing the world. Report, PwC, London.

Winkler, T. (2011). Retracking Johannesburg: Spaces for participation and policy making. *Journal of Planning Education and Research*, 31(3), 258–271.

Yigitcanlar, T., Han, H., Kamruzzaman, M., Ioppolo, G., Sabatini-Marques, J. (2019). The making of smart cities: Are Songdo, Masdar, Amsterdam, San Francisco and Brisbane the best we could build? *Land Use Policy*, 88, 104187.

4

Reassuring

4.1. Introduction: a transparent city, from reality to illusion

Within smart cities, digital infrastructures have often developed "under the radar", invisible to everyone, without citizens or sometimes even public decision-makers being fully aware. However, the time has come to think about the digital relations between public authorities and citizens. For example, in 2023, the city of Angers created a participatory democracy body, the local digital council, engaging 40 citizens (20 men and 20 women from different neighborhoods) in discussions on digital technology. Another example is the Montpellier metropolis, which after banning facial recognition in October 2022, set up a local convention on artificial intelligence (AI) in 2023, allowing 40 citizens drawn at random to reflect on the uses of AI. The time therefore seems to have come for citizen consultation and open exchanges on digital developments and applications in the field of public action.

Transparency around smart city practices and technologies thus appears both as a moral imperative, essential for restoring trust between public authorities, elected officials and citizens, and as a strategy aimed at legitimizing the deployment of technologies and processes (sensors, the Internet of Things, digital channels, AI, algorithms, etc.) within the public space. For example, the Angers metropolitan area has set up pictograms in the city, conforming to the DTPR standard for "digital transparency in the public domain" (under free license), which "can provide information on the objective (enabling Internet connection, reducing energy consumption), on the type of technology (contactless payment, de-identified video, i.e. video processed in such a way as to remove identifying characteristics before it is used or stored), the type of data collected, and their storage" (see below for the visuals sent by Jacques Priol, as well as his box in Chapter 5). However, the notion of transparency is rarely defined or questioned.

Chapter written by Audrey PORTES and Gilles N'GOALA.

Figure 4.1. *The case of Angers metropolitan area: Digital Trust for Places and Routines (DTPR). For a color version of this figure, see www.iste.co.uk/ngoala/smartcity.zip*

Is transparency the art of being visible or the art of being invisible? While for engineering sciences, a transparent technology is one that is not apparent to users so as to deliver fluid experiences (e.g. a smartphone whose intrinsic functioning is invisible); in the human and social sciences, it is generally seen inversely, as practices and usages that should be revealed and made visible to the various stakeholders (citizens, media, elected officials, etc.). For smart city stakeholders, should transparency then lead to technologies being made visible at the risk of complicating the experience and degrading the services provided to citizens? Or is it about revealing digital practices and uses within the public space while making the sensors, applications, technologies, data and metrics collected by public authorities and businesses invisible? While transparency is often presented as a moral imperative in our democratic and open societies, the question of what information can, must, or should be disclosed to citizens on the practices and uses of the smart city, its metrics, its technologies, its usefulness, its limits, etc., does not lead to an obvious and definitive binary answer (yes or no).

Indeed, how can we reconcile invisibility of technologies and transparency of practices in the management of the smart city? Would transparency alone be enough to restore trust in digital technology and smart city projects? Faced with the misunderstandings which the digital transformation can evoke, does total transparency not risk creating generalized suspicion and disproportionate mistrust?

In terms of technology, is not transparency an illusion, an element of language, a reassuring discourse for citizens, about technologies which present a form of opacity to its users (for example, the black box of AI)? Is it not, on the contrary, necessary to admit that the smart city cannot be totally transparent, provided that its users make controlled, benevolent and honest use of it? Is not trust in institutions a prerequisite – and not a consequence – of transparency as perceived by citizens? Is a citizen demand for transparency not symptomatic of distrust toward institutions and elected officials? After having defined what a transparent smart city is, we question its links with the establishment of a society of trust and the opportunity to give citizens the power to control and influence the construction of a digital city.

4.2. Definition of a transparent smart city

The word "transparency" is of Latin origins, coming from *trans*, meaning "beyond" or "through", and from *parere* meaning "to appear" or "to show oneself". Originally, it meant "showing light through", but its meaning evolved to "easily seen through", implying the presence of a medium that fades away to reveal what lies behind. From a geological point of view, transparency is defined in terms of the amount of light that can pass through a material. This definition is metaphorically linked to the transfer of information in a relationship. Different levels of transparency can be conceived along a continuum ranging from opacity (non-transparency) to clarity (transparency) through translucency (an attempt at transparency, but with biases). This metaphor of light highlights that there are various levels or degrees of transparency, which can be considered as strategies for stakeholders within the smart city.

4.2.1. *Smart city and transparency*

According to Martinez (2001, p. 40), "transparency does not describe an action, but a state of perfect communication between the spectator and the phenomenon that appears to the spectator despite the obstacle". In management, transparency may take different forms and, according to Hultman and Axelsson (2007), refers to four distinct objects: transparency of costs, the organization, the supply chain and technology. Granados et al. (2010) also identify several objects of transparency in a B2C market, including prices, product information, availability of inventory, costs of production and distribution and internal transaction and management processes.

Transparency and the smart city: is the adjective very appropriate or contradictory to the concept?

By Philippe Sajhau, Director of Smart Cities, Innovation and Data, Noisy-le-Grand (France), former Vice-President of IBM France for smarter cities and metropolitan programs.

The smart city has developed a lot in almost 15 years, moving from "all technology" to a vision closer to residents, where technology is only a tool at their service and that of local agents to provide better services and help support the ecological transition. While the question of low tech is very present today and deserves to be considered, the fact remains that the smart city is mainly based on the use of data, whatever its source. It is an important asset for a city or territory to support public decision-making, build services or provide information to the population. The question of transparency arises in this context, since the word "data" is commonly associated with a lack of respect for privacy, its abundant use by international operators who abuse data that are nevertheless created locally, or even with "tracking" an activity or a person. What is it, really?

Let us first consider the smart city as the result of the use of data: air sensors in public spaces to measure its quality, presence sensors in rooms for better control of heating, multimodal travel applications allowing users to choose their mode of travel and detect traffic density, video cameras to count mobility flows or provide video surveillance for residents, personal data collected by various city services (civil status, culture, sport, etc.). I could continue to name examples which involve collecting information with (or without) human presence on the public street. The data are then processed on servers (located in France or elsewhere), and then most often reconstituted on a dashboard or a mobile application. During these operations, the risks of "violating" the privacy of residents are minimal under the so-called GDPR law (General Data Protection Regulation), and any error or desire to ignore it is punishable by heavy financial penalties. The city must nevertheless ensure that it alerts residents that it is collecting this information and ensures that it communicates how and for what purpose it processes this information. Many cities have deployed data charters which define a guide to good practices and commitments on the elements of transparency or sovereignty of their digital action. In Noisy-le-Grand, we envisage doing this, but also opening a DATALAB to the population, a sort of open place to share the uses of data, promote digital inclusion and demystify the word "data". Along the same lines, we are launching an initiative with a start-up called Kesk'IA to talk to young people about AI and have them develop algorithms to demonstrate simple everyday uses. These are initiatives that must support the development of new uses through data to inform the population about the desired objectives and help them better understand what is being done by the local authority and thus reassure them.

However, the smart city also allows us to touch on another aspect of the subject of transparency: the transparency of public action and public participation in its choices.

The concept of Open Data is now widespread (and, moreover, the law for municipalities with more than 3,500 inhabitants). The aim is to provide residents and economic stakeholders with sets of data that explain the activity and public action of the city or territory. Originally, the aim was to ensure transparency in public action, with information and data easily accessible to residents. The state itself is today a major provider of open data: just one example is the DVF site of ETALAB which publishes the latest real estate transactions for a city or a neighborhood over the last three years, a useful tool even for a city's urban planning services, but especially for those looking to acquire housing. Noisy-le-Grand opened its site in March 2022, with 37 data sets, and many territories are moving toward publishing this activity online alongside the city's other communication tools.

The other means of openness and transparency available to municipalities is a consultation application, part of the range of tools called CIVITECH, also including participatory platforms and applications allowing problems on public roads to be reported: potholes or fly-tipping, so that the town hall can act as quickly as possible. The consultation tool is a way for elected officials to get opinions from the population (in addition to physical meetings) on what they want, so that they can contribute to decision-making. In Noisy-le-Grand, we developed a tool called DECIDIM which allowed us to consult the population on major issues, such as the place of trees in the city, but also to allow them to propose projects within the framework of the participatory budget and then to vote transparently to retain those that were most in demand and in compliance with the proposed budget.

The smart city, which today I prefer to call the "intelligent", "innovative" and "sustainable" city, may evoke fears of "opacity" in the use of data and its processing. It is therefore necessary to ensure that explanations, information and places for sharing on the subject are organized to reduce this fear. However, at the same time, it represents a real opportunity to share the content of public action more simply to open it to all in a simple and permanent way. It also allows co-creation with parts of the population who were not accustomed to dialog with elected officials. From a risk to transparency, the smart city might come to be seen as an opportunity for transparency.

Box 4.1. *Transparency and the smart city*

In terms of the smart city, the issue mainly relates to the use and exploitation of individual or aggregated data by public authorities or other actors: digital giants, utility companies operating in cities (energy, water, waste, transport, etc.), small businesses, associations, etc. Digital transparency (Portes et al. 2020) questions in particular the possibilities that remain for citizens to access their data, to understand why and how they are used, and to control and decide whether to disclose them to third parties or not (Awad and Krishnan 2006; Martin et al. 2017). Even if the term "smart city" is not limited to the idea of a digital city – it also concerns sustainability and soft technologies – it is the digital transformation of cities and the massive collection and exploitation of data that raise concerns and test citizens' trust in public authorities (see testimony below from Philippe Sajhau, Director of Smart City, Innovation and Data, Noisy-le-Grand, France).

4.2.2. *Digital transparency at the heart of a new social contract between citizens and decision-makers*

Depending on the context, the data which are collected do not involve the same risks in terms of privacy and personal freedom, for example, weather data are not the same as facial or voice recognition data. However, as can happen in non-democratic states, the inclusion of digital technology in the public space does indeed raise fears of the emergence of a society of surveillance, control and "steering" by public authorities and the possible manipulation of citizens for economic or political ends. The Chinese government's social credit project, for example, leads citizens being rated and "scored" based on personal data collected, using millions of video surveillance cameras coupled with facial recognition, geolocation, exploitation of purchasing data, social or medical networks, etc. The issue of order, security and tranquility thus often prevails over the principles of individual freedom.

In democracies, the law guarantees citizens a form of protection against potential abuses of digital technology. For example, the General Data Protection Regulation (GDPR), implemented in 2018 in Europe, provides some answers. It represents a significant change in the culture of companies and institutions regarding data practices by imposing new ethical, responsible and transparent requirements.

Rousseau, in his book *The Social Contract*, in 1762, advocated transparency as a fundamental principle of political organization, emphasizing the disclosure of information through standardized reports concerning various aspects, such as the headquarters, history or evolution of the company. As a principle, transparency assumes the status of a norm, that is, an injunctive (what the public authorities must do) or descriptive (what the public authorities usually do) expectation by the citizens. Transparency is thus enshrined in freedom of information laws that oblige governments to disclose information to citizens (the functions of the CNIL – French National Commission for Information Technology and Civil Liberties, GDPR, Digital Services Act, etc.). It is thus presented as a fundamental human right, the "right to know" and to consent to the use of somebody's personal data.

Transparency is also part of a relationship of exchange between different stakeholders (citizens–elected officials, customers–companies, etc.), each of which must determine how much, how, where and when to reveal information. Information asymmetry is a phenomenon inherent in every exchange, with each party having a different level of knowledge and information. For example, a developer of AI solutions cannot always explain the complexity of its tools and algorithms to its users. This black box phenomenon and the inexplicability of many AI solutions drastically reduces the possibilities of resolving information asymmetries and potentially opens the way to opportunistic behavior on the part of "those who know".

In economics, Akerlof (1970), in his famous article *The market for "lemons"*, emphasizes both the importance and the difficulty of reducing information asymmetry to balance the markets and avoid the dominance of only bad products (the phenomenon of adverse selection; the term *lemons* indicating bad used cars in the United States). All economic agents are in fact caught in a dilemma between their willingness to cooperate, which would lead them to free up information, and the quest for their own personal interest, which would lead them to exploit asymmetry for their own benefit and to exercise more power during the transaction. There appears to be a gap between what individuals want to know and what other entities want to disclose, because this arises from the power relationship between the partners.

To resolve information asymmetries, two complementary elements prove essential for transparency in the exchange.

– Reciprocity: through exchange, individuals seek to maximize their satisfaction while minimizing their costs, implementing strategies to balance their desire for cooperation and exploitation of asymmetry. The theory of social exchange thus promotes the idea of reciprocity in exchanges, which implies knowing how to give to receive information within a transaction mechanism.

– Equity: the theory of equity emphasizes the importance of seeking balance in the exchanges between the parties. When individuals feel that they are giving more than they are receiving, this can lead to negative reactions (complaints, negative word of mouth, retaliatory actions, etc.).

Information asymmetry is inherent in relationships of exchange, but requires, on the one hand, ensuring ethical practices in the use of data, in accordance with society's norms; and, on the other hand, clarifying the terms of the exchange. In other words, the smart city constitutes a legitimate project on condition that it complies with norms, which vary according to countries and cultures, and demonstrates the gains and benefits gained by the residents. The collection of data by city stakeholders (local authority, delegated companies, etc.) without any economic, social or ecological compensation, whether on an individual or collective level, would be unfair. As happened in France with the Linky meter, obtaining metrics and sharing data must be justified by new services, generating savings (energy, water, etc.), and/or contributing to individual or collective well-being. Otherwise, it might lead to social rejection from the population.

4.3. Transparency and trust: the causalities

In a digital and connected city, transparency is sometimes presented as an insurmountable moral principle, sometimes as a stakeholder's policy (local authority, company, etc.) in the context of exchanges with other stakeholders (citizens, consumers, etc.). However, transparency cannot be fully translated into a law, a regulation – including the GDPR – or a contract that is by nature incomplete, or exhaustively consider all of the information that might be necessary to reduce information asymmetries and balance power relations between citizens on the one hand and digital stakeholders on the other (local authorities, digital giants, start-ups, etc.). Contrary to the opportunistic logic that may prevail in certain economic circles, transparency also relies on the goodwill of stakeholders and their desire to establish reciprocal, fair, honest and good faith exchanges. Transparency and trust are thus intimately linked.

4.3.1. *From transparency to the establishment of a digital society of trust: a complex path*

The issue of transparency is nothing new in the fields of political science, management and public communication. It refers first to the idea of "revealing yourself" to the view of citizens. The latter are particularly called upon to look through the "windows" of an institution or government (den Boer 1998) and to become external spectators of a top-down flow of information (Ruppert 2012). Institutions can thus establish a form of citizen oversight of public affairs (Birkinshaw 2006). It also refers to an objective and neutral way of delivering information (Rawlins 2009), even when it means having to assume our mistakes and responsibilities during crises and critical situations. Finally, it refers to a greater openness of institutions toward citizens. Transparency is also built in a dynamic and collaborative way through interactions between public authorities and citizens, for example, through consultations, participatory bodies or neighborhood committees.

Transparency cannot be considered solely as a static phenomenon consisting of unilateral provision of complete and relevant information to stakeholders. In a more dynamic form, transparency involves continuous interaction between its stakeholders and relies on greater interactivity between the parties, personalization of information for each party and the establishment of an open conversation. Ultimately, transparency can be seen equally as a norm, a legal or ethical obligation or a subjective perception that depends on individual skills, knowledge and expectations regarding information (perceived transparency). It may be static, consisting of disclosure of information by one party, or dynamic, involving ongoing interactions between stakeholders.

According to Portes (2018), digital transparency is therefore based on three main dimensions or levers.

– Limpidity of information: this refers to the clarity and relevance of the information disclosed. Simple disclosure of information does not guarantee transparency. Information must be comprehensible, visible, accessible and contextualized to be considered transparent. It is essential for information to be disseminated at the right time to promote transparency.

– Objectivity of information: this dimension implies truth, honesty, frankness and absence of dissimulation. A transparent organization hides neither its good actions nor its bad actions. Objectivity also means the capacity to disclose negative information about yourself, which reinforces perceptions of transparency.

– Openness of information: this dimension emphasizes two-way communication and dialog with stakeholders. Transparency is not limited to the unilateral disclosure of information, but also involves creating an open dialog and facilitating communication between the organization and its stakeholders.

In summary, transparency is a multidimensional concept and is not limited solely to disclosure of information but also encompasses how this information is presented, understood and shared with others (Portes et al. 2017).

Transparency is of fundamental importance in building a digital society of trust. Without transparency around the digital uses and practices of public authorities and other stakeholders (companies, experts, etc.), trust seems impossible and the engagement of citizens in public life seems futile. Trust in digital technology becomes all the more essential as citizens face large information asymmetries (digital giants master this field) and see themselves as dependent (it is difficult for them to do without digital technology) and vulnerable due to their low level of competence and their potential powerlessness in the face of technologies beyond their control (hacking, data theft, identity theft, intrusion into privacy and disclosure of private data, etc.).

The relationship between transparency and trust, however, is more complex than it seems. For Portes et al. (2020), the objectivity of the information certainly contributes to strong reinforcement of trust, that is, the belief that the partner (company, public authorities, etc.) is credible, honest and benevolent toward individuals. Transparency is thus a means of restoring trust by reducing information asymmetry in digital environments, such as that of the smart and connected city. However, excessive limpidity about digital uses and practices (collection, use, sharing of data, etc.) may, beyond a certain threshold, create fear and suspicion, whether founded in fact or not, and thus arouse mistrust. Each digital stakeholder must therefore strive to find a median and optimal level of information limpidity, that is, saying neither everything nor nothing about their digital practices, to optimize the trust that individuals have in them. Furthermore, opening up information to individuals does not help to strengthen their trust, but does manage to strengthen their engagement, for example, through participation in events and workshops, word-of-mouth transmission or the production of various forms of content useful to the organization (suggestions, ideas, innovation, etc.). In short, trust and transparency are key elements in developing exchanges and relationships within smart cities. Transparency may certainly contribute to trust, but it must be managed in a balanced way to avoid undesirable effects.

With the development of a more horizontal society due to the development of social media, transparency has gradually become an argument for communication in the public and private spheres. Many regularly proclaim their transparency and develop communication and persuasion strategies around this value, without always being transparent about what it means in concrete terms or what actions are carried out beyond simply respecting the laws and regulations imposing transparency (use of data, explicit consent, geographical origin, costs and prices, etc.). To be credible, transparency must be established more through evidence (limpidity, objectivity and openness of communication) than through an agreed narrative responding to the injunctions of citizens and multiple stakeholders.

4.3.2. *The illusion of transparency: between citizen trust and distrust*

To create a society of trust, particularly toward elected officials and public decision-makers, transparency is often presented as a prerequisite. It is widely accepted that a lack of transparency can lead to behaviors of distrust or resistance regarding the decisions made. For example, an organization that is non-transparent about its data collection practices exposes itself to citizen/consumer actions such as installing anti-spam software, sharing false personal information and creating multiple online identities. Furthermore, as shown in political communication, in the digital age, more knowledgeable and better-informed individuals may ultimately be more distrustful (Gallot and Verlaet 2016).

The injunction to transparency upheld by associations, citizens, the media and political figures is however not unanimously recognized by all observers. Many cynically believe that transparency relates to a digital utopia (Jeanneret 2005) or an illusion that would legitimize the place of digital technology in the public space. The transparency imposed at present as a necessity and even as a rule of law in the commercial and social sphere implies the idea of sharing a lot of information of showing yourself by participating in the "transparency game" or the "digital society game" (Gallot and Verlaet 2016). According to Onfray (2001), this "game" involves a strategy, an intentionality, and a staging on the part of stakeholders who deliver a truthful message without revealing the whole truth, all with the aim of protecting their own interests. Transparency is thus assimilated to a strategy, or even to a manipulation of the other party in the exchange. Indeed, how could "ordinary" citizens understand sensors (IoT) and digital applications, algorithms and AI, when these devices often resemble impenetrable and indecipherable "black boxes"?

Transparency may be at the same time a synonym for truth and an instrument of manipulation. It may sometimes hide more than it reveals, because it may be selective in disclosure of information. The strategy of transparency consists of showing what is true, but not necessarily the whole truth. Assessing whether transparency efforts are motivated by manipulative intentions remains difficult, as no one will admit to having such intentions. On the other hand, citizens may also lack transparency toward stakeholders in the smart city. They may be torn between the need to adhere to the system and the desire to protect themselves while enjoying its benefits. This can sometimes lead to opportunistic or deviant behavior (fraud, lies, etc.), which are failures to obey group norms.

So, should we regret that transparency is sometimes only an illusion? Probably not. Secrecy is, paradoxically, consubstantial with the notion of transparency. Total transparency about everyone and everything cannot reasonably exist in society. Its limit is found in the secrecy that a society must ensure for its members (individuals, organizations, etc.). Every citizen's personal information should not be made public, nor should the trade secrets of companies be shared with competitors or other countries. This follows from individual freedoms and state sovereignty.

The choice is therefore not whether to be transparent or not, but to determine what we should let show and what we should keep secret. For example, in Finland, in the name of transparency, the tax authorities publish the taxable income of each citizen every year in November, allowing everyone to know what their neighbor, their ex-spouse, their children, the country's celebrities, etc., earn. However, in France, such a practice of tax transparency would probably be considered stigmatizing, voyeuristic and morally unacceptable. It would especially accentuate the social control exercised by some over others and would potentially generate a society of jealousies. On a political level, transparency, which is often presented as an asset for democracy, could on the contrary lead to a totalitarian state which would establish extreme control over everyone, who would then be required to tell it everything and to tell everyone everything. It would therefore be better a priori to have an illusion of transparency than a dictatorship of transparency, which would be imposed on everyone, causing harm to some and restricting the individual freedoms of others (de Fabrègues and Pernod-Burckel 2017).

Like all societies, Finland does not disclose all information about its citizens and keeps the essentials secret, to preserve the privacy of each and the individual freedoms of all. Thus, transparency is a relative value, specific to a culture, a context or a person, and not an absolute moral value. It is evaluated according to the context in which it is exercised and the social acceptability of the transparency practices in question. It leads to the illumination of certain realities, while certain more intimate, less confessional and more secret truths are pushed into the shadows. This is the case for the smart city. Selection of information on the digital uses and practices of the main stakeholders (communities, digital companies, etc.) is necessary. It must also be carried out in consultation with interested and concerned citizens with constant attention to preserving the interests of multiple stakeholders.

4.4. Empowerment of citizens as a condition for true digital transparency

To commit to transparency, for example, by sharing information with citizens on the uses of AI and data in the city (see the example of Angers with the DTPR pictograms), it would still be necessary to ensure that citizens have the capacity to understand, master and control their digital environment and to act individually or collectively to preserve their private life and their intimacy. It is by supporting citizens toward digital technology and giving them more power (empowerment) and capacities that it will become possible to make the smart city a more transparent object.

Morrongiello et al. (2017) notably distinguish the power of choice (power to), relative to the increased cognitive and behavioral capacities of individuals (information, mastery, control, etc.), power over others (power on), relating to each person's capacity to influence their surroundings or anonymous people via social media, for example; and finally, the power to act together (power with), relating to the capacity to integrate into collectives in order to have a greater impact on their digital and sociopolitical environment. The increased power over the digital society that is offered to citizens could then be more of an opportunity than a constraint, if everyone "plays the game" and accepts the rules.

4.4.1. *Empowering citizens to act and to control the digital public space*

Digital technology has often gradually imposed itself on citizens without them being able to decide. For example, access to many public services now requires citizens to master digital channels and tools (searching for information, bank transfers, etc.). The level of constraint, however, varies depending on the situation.

Three categories of spaces have been identified based on the level of constraint that they impose on individuals:

– The space of suggestion offers many possibilities and a range of behaviors to individuals, allowing them a high degree of appropriation and control. For example, they may choose to use digital or physical channels to access services.

– The space of seduction influences individual practices by directing them toward certain actions. It offers a medium level of appropriation and control. For example, campaigns may be organized to raise awareness and encourage digital uses to replace physical procedures that consume time and effort.

– The space of prescription forces individuals to follow a specific behavior defined by the organization. In this case, there is very little potential for appropriation and control by individuals. For example, the forced march toward dematerialization and the use of digital channels (mobile applications, etc.) has become prevalent in certain areas (90% of French people file their taxes online, for example), with non-users being systematically penalized in case of default (difficulty of access, queues, additional costs, etc.).

The concept of these three types of spaces helps us to understand the relationship that an organization can establish with citizens, depending on the degree of constraint or interpretative flexibility that it offers in the environment.

Type of relationship	Definition	Level of citizen control
Suggestion	Offers a plurality of possibilities, a range of behaviors	Strong
Seduction	Influences practices by inviting, leading, or directing the citizen toward certain practices	Medium
Prescription	Requires and forces the citizen to act according to a single type of behavior planned by the organization (*blueprint*)	Weak

Table 4.1. *Gumuchian's (1991) levels of spatial order adapted to the relationship within the smart city*

Citizens are looking for efficiency and can, through using digital technology, save time, make less effort, make their lives easier and access numerous resources (information, services, etc.). This search for efficiency involves control over their environment, defined as "a structuring force of activity oriented towards the need to demonstrate one's competence, superiority and mastery of the environment" (White 1959).

The challenge is therefore to make the digital space appropriable and controllable by users. However, according to environmental psychology, appropriation depends, on the one hand, on the individual exercising mental or physical control over the space; and, on the other hand, on the room for maneuver offered by the context and the organization of the space, which represents a system of constraints. There must be "interpretive flexibility" in the digital space to allow appropriation, that is, opportunities must be left for individuals to exercise some control.

In a smart city, the question therefore arises as to what freedom should be left to citizens to facilitate their appropriation of digital technology and increase their ability to master it, or even divert it from its prescribed uses (adaptation, creation, psychological possession) (Mifsud et al. 2019). Empowerment of citizens therefore assumes that the public and private organizations in charge of the smart city agree to share control of the digital public space with other stakeholders, and with citizens who aspire to regain control over a digital environment that is beyond their control.

Between the empowerment of citizens and their alienation from digital technology (powerlessness), elected officials and public decision-makers must now establish a vision, a strategy and a roadmap. They certainly must face many citizens who adopt pessimistic and avoidant attitudes toward digitalized environments. These latter feel powerless and resigned as they observe the galloping evolution of digital technologies (AI, connected objects, etc.), while being anxious and concerned about risks linked to their private lives and their individual freedoms. Others, on the other hand, aspire to a greater level of control, power and mastery, even domination, over their digital public space. With these, elected officials and public decision-makers must engage in conversation and agree to cede power to them, and give them the means to exert it. This involves ongoing assistance and support with new digital learning.

4.4.2. Empowering citizens to understand digital technology in the city

In the context of the smart city, individuals demonstrate varying levels of competence. Competence, according to Le Boterf (1994), is not limited to knowledge or know-how, but is manifested in the act of mobilizing diverse resources to solve specific problems. Three types of competences are identified:

– *Cognitive competences*: these involve the ability to process information effectively, including the ability to decode discourse, search for, understand, evaluate and organize information in a relevant way.

– *Instrumental competences*: these competences are related to the coordinated mastery of tools, techniques or technologies to accomplish specific tasks.

– *Competences of use*: they are directly related to the use of products, services or media, and involve the ability to use them appropriately.

In the digital context, "digital literacy" refers to the ability to understand and master the Internet, digital tools and information technologies. It allows individuals to use digital technology in their daily lives, at work and in society to achieve personal goals and broaden their competences. "Digital literacy" is not limited to theoretical knowledge, but also includes the ability to apply this knowledge in practical ways, to interpret, create and communicate information in an ever-changing digital environment. It is essential for developing effectively in the digital world of the smart city.

Within the smart city, some individuals go beyond the mere use of technology. They seek to understand their environment in depth to gain significant benefits from it. These individuals are not content to accumulate knowledge about their environment but invest time and effort to exploit its opportunities more effectively. This ability to decode and exploit the environment is often called "insight".

Insight is characterized by the ability to profoundly understand our market or environment, evaluate trends and opportunities, and implement a strategy to attain specific results. It encompasses several dimensions, including the ability to analyze the current environment, to understand the overall functioning of the market, to establish connections between actions and results, to have a global vision of the situation, to be constantly in search of knowledge to reduce ambiguity and to exercise effective leadership. Insightful individuals are willing to take on challenges, ask pertinent questions, exercise keen discernment when making decisions and adopt a critical attitude toward their environment. Insight is not innate, but a competence acquired over time through learning and experience. In summary, insightful individuals within the smart city are characterized by their ability to understand their environment in depth, assess opportunities and challenges, and make informed decisions to make the most of it.

Elected officials and public decision-makers are thus faced with a very strong heterogeneity among citizens in terms of competences and approaches to digital technology. Believing that the smart city can be co-designed, thought out and implemented with most citizens is certainly an illusion. However, instead of designing the smart city for users/citizens, it is important to do it with some among them, who are more inclined toward innovation (technological, ecological, social, digital), and to make more use of their ideas, creativity, feedback and imagination. We can then ask ourselves whether the structures established for participatory democracy (neighborhood committees, development councils, etc.) fully correspond to the objectives sought for a more innovative city.

4.4.3. *Empowering citizens to act collectively to preserve their privacy and individual freedoms*

The smart city refers to both a utopian imaginary (a quality of life ensured by technology) and a dystopian one (human beings dominated by machines and artifacts). It is fundamentally associated with the idea of an innovative city, where technological progress has a positive impact (tech for good, smart green tech, etc.) and benefits the greatest number. However, this initial postulate is not unanimous, with some advocating, on the contrary, a detechnologization of the world (N'Goala and Collin-Lachaud 2023). Beyond doubts about the possibility of reconciling technological development and environmental preservation, the race toward digital technology raises serious questions for governance and digital sovereignty. Data have certainly become an economic issue, the "new oil" of the platform economy (the GAFAM companies, in particular), but also a sociopolitical issue for citizens concerned with preserving their freedom and control over their personal data. The opposition to smart city projects, for example, in relation to hypervisors which film the streets and allow facial recognition to be implemented, as at the Paris 2024 Olympic Games, mostly partakes in this sociopolitical perspective. Collective movements (associations, parties, etc.) are thus set up to influence public decisions (laws, regulations) and support the establishment of good practices.

However, it is important to note that the digital public space is not dominated by public actors alone, quite the contrary. Large private platforms (Amazon, Google, Apple, Airbnb, WeChat, etc.) dominate this world of digital and AI, and operate within the physical spaces of territories. For example, each order on Amazon generates flows of information (exploited data) and merchandise (urban logistics) that are largely beyond the control of public authorities. Users/consumers/citizens may have become more accustomed to providing their personal data to these private actors than they are willing to accept for public authorities. This may seem paradoxical, given the lack of moral and political legitimacy of these private actors. On the contrary, this may seem natural to citizens who fear political and social control of their behavior in the public sphere. The intentionality inferred by citizens (earning money for private companies versus controlling people's lives for public authorities) might explain citizens' responses.

Citizens oscillate between resignation in the face of a digital world that is beyond their control and a desire to better control their personal data. Resignation is the feeling of helplessness and passivity, as opposed to the need for control. Learned helplessness theory, developed by Seligman and Maier (1967), explains how individuals exposed to uncontrollable situations may develop resignation behavior. When individuals feel that their efforts to understand their environment are in vain, this may lead to feelings of helplessness and resignation.

The concept of resignation is related to perceived helplessness, where individuals believe that they cannot achieve positive outcomes through their own actions. This may be perceived as a form of alienation, where individuals feel disconnected and helpless in the face of external forces (business, technology, AI, etc.). Resignation has negative consequences, including dissatisfaction and mistrust. Organizations have an interest in understanding the mechanisms of alienation to struggle against this type of perception. Helplessness can be defined in two ways: powerlessness, where a person cannot change the outcomes of a specific situation, and learned helplessness, which occurs after repeated failures to control the outcomes of an action and can affect a person's belief in their ability to exercise control in other, unrelated situations.

Resignation occurs after repeated exposure to deprivation of control or uncontrollability, leading to decreased performance in carrying out tasks. Learned helplessness occurs when individuals have already had a bad experience in each situation and do not expect their actions to change the outcome, leading to resignation behavior. Resigned individuals may be less inclined to act against the smart city, as they feel limited in their actions due to their powerlessness. However, there may also be psychological reactance, where they attempt to regain a lost sense of power. Resigned individuals are more vigilant, but they may also feel negative emotions and have unfavorable attitudes.

Conversely, other citizens may wish to regain control over their personal data and avoid the risks of intrusion into their privacy, dissemination to third parties and *hacking* (identity theft, ransom, etc.). Institutions (for example, the National Commission for Information Technology and Civil Liberties in France) and laws (for example, the General Data Protection Regulation in Europe) certainly aim to protect users and citizens, with local authorities, for example, being obliged to have a data protection officer (DPO). However, some citizens, considering themselves individually and collectively vulnerable, do not believe that laws and institutions provide them with sufficient guarantees and act to establish a more protective legislative and regulatory framework (moral legitimacy) or to obtain greater benefits in exchange for their data (pragmatic legitimacy).

From a perspective of exchange, users in a city may accept the collection of their data, but also wish to be remunerated and obtain compensation for the value of the data they provide (services, savings, etc.). In return for collecting data, the city must particularly provide better services (mobility services, waste management, etc.), facilitate resource savings (energy, water, lighting, etc.) and ensure a better quality of life (security, health, etc.). When individuals provide information, they establish an implicit social contract with the organization collecting the data. This trade-off is based on a comparison between the value of using personal data and the potential negative consequences of disclosing this information. In this exchange perspective, to accept this exchange relationship with an organization, the individual must perceive that the benefits clearly outweigh the costs and risks.

The objective of transparency might lead to explaining what data are collected and how it is collected, then clarifying what this will enable to be provided, such as simplified procedures, reduced waiting times or personalized offers.

It is necessary to foreground the sharing of the value of data between the different stakeholders of the smart city. Companies should value reciprocity in the exchange of data rather than simplifying their discourse by claiming that data are collected only to improve experience. Citizens must be able to judge whether the exchange of data for services is fair.

From a legal perspective, users in a city can also plead for more guarantees regarding:

– Their right to information: the right to know how the data are collected and used, as well as the benefits that will be obtained in exchange for collection and use of the data.

– Their right to control: the right to act on the data, to choose whether to disclose their data or not, and to decide on its use by specific entities at certain frequencies. This implies the possibility of giving our opinion on the dissemination of data.

– Their right to anonymity: the development of AI for facial and voice recognition raises the risk of loss of anonymity in public spaces. In addition to these tools, algorithms can also combine data to reconstruct user profiles. The public space then ceases to be a space where everyone can enjoy a form of anonymity.

These two perspectives constitute two ways of legitimizing the smart city. In democratic societies, citizens may accept the implementation of data collection and exploitation, on condition that they can assess the advantages and be assured that their rights will be respected. Otherwise, public and private decision-makers may face strong resistance.

4.5. Conclusion

Transparency and the smart city may seem like a perfect oxymoron, that is, a juxtaposition of two contradictory notions. Indeed, digital technology in the public space potentially increases the information asymmetries between organizations (states, communities, businesses, etc.) and citizens. On the one hand, citizens are becoming increasingly transparent to organizations, especially since they can be easily identified and monitored anytime, anywhere and on any device. On the other hand, the tools, practices and technologies used are generally invisible and remain opaque from the users' point of view (data, algorithms and AI). Faced with the acceleration of the digital transformation, most citizens are unaware of the uses that will be made of their data.

Faced with the growing opacity of the digital world, the massive collection and use of data, and the increasing use of AI, the need for transparency is becoming increasingly important. Citizens evolve with their digital environment, acquiring new knowledge and competences, even digital ease. However, they also feel a loss of control with respect to an extremely changing and uncertain environment. This duality of attitudes and perceptions can be compared to the distinction in psychology between approach motivation (seeking positive stimuli) and avoidance motivation (moving away from negative stimuli), both of which are necessary to respond effectively to the environment.

Approach attitudes in a connected environment reflect individuals' desire to better understand organizations and their practices, encouraging them to play an active role in their adaptation to the digital environment. Some individuals possess a digital ease that allows them to navigate more easily in this environment because of their expertise and their digital behaviors. Individuals therefore react in different ways to the evolution of the smart city and the digital world, oscillating between optimism linked to mastery of this environment and anxiety about loss of control and risks.

In order not to be illusory, transparency must include the empowerment and digital education of citizens so that they understand the changes in their environment and master the associated digital practices. This may be achieved through initiatives such as tutorials on personal data management and digital education programs (see the CNIL's initiatives on this subject). Authorities must play a proactive role in educating citizens, helping them to understand and master their digital environment. This will help avoid stress and anxiety among residents, reinforce trust and engagement in these digital environments and make the goal of transparency less illusory.

4.6. References

Akerlof, G. (1970). The market for "lemons": Quality uncertainty and the market mechanism. *Quarterly Journal of Economics*, 84(3), 488–500.

Awad, N.F. and Krishnan, M.S. (2006). The personalization privacy paradox: An empirical evaluation of information transparency and the willingness to be profiled online for personalization. *MIS Quarterly*, 30(1), 13–28.

Birkinshaw, P. (2006). Transparency as a human right. In *Transparency: The Key to Better Governance?*, Hood, C. and Heald, D. (eds). Oxford University Press, Oxford.

den Boer, M. (1998). Steamy windows: Transparency and openness in justice and home affairs. In *Openness and Transparency in the European Union*, Deckmyn, V. and Thomson, I. (eds). European Institute of Public Administration, Maastricht.

de Fabrègues, M. and Pernot-Burckel, V. (2017). La dictature de la transparence. Patrimoine et éthique publique : enjeux politiques et professionnels de la représentation et de la communication du patrimoine. *Zone libre*, 19(2).

Gallot, S. and Verlaet, L. (2016). La transparence : l'utopie du numérique ? *Communication & Organisation*, 49(1), 203–217.

Granados, N., Gupta, A., Kauffman, R.J. (2010). Information transparency in business-to-consumer markets: Concepts, framework, and research agenda. *Information Systems Research*, 21(2), 207–226.

Gumuchian, H. (1991). *Représentation et aménagement du territoire*. Economica, Paris.

Hultman, J. and Axelsson, B. (2007). Towards a typology of transparency for marketing management research. *Industrial Marketing Management*, 36(5), 627–635.

Jeanneret, Y. (2005). La "société de l'information" comme figure imposée. Sur un usage particulier des mots en politique. In *La "société de l'information", entre mythes et réalités*, Mathien, M. (ed.). Bruylant, Brussels.

Le Boterf, G. (1994). *De la compétence. Essai sur un attracteur étrange*. Editions d'Organisation, Paris.

Martin, K.D., Borah, A., Palmatier, R.W. (2017). Data privacy: Effects on customer and firm performance. *Journal of Marketing*, 81(1), 36–58.

Martinez, J.P. (2001). Dossier : sûreté nucléaire et transparence [Online]. Available at: https://www.asn.fr/Media/Files/surete141.

Mifsud, M., Matthieu, M., Cases, A.-S., N'Goala, G. (2019). It's MY health care program: Enhancing patient adherence through psychological ownership. *Social Science & Medicine*, 232, 307–315.

Morrongiello, C., N'Goala, G., Kreziak, D. (2017). Customer psychological empowerment as a critical source of customer engagement. *International Studies of Management & Organization*, 47(1), 61–87.

N'Goala, G. (2015). Opportunism, transparency, manipulation, deception and exploitation of customers' vulnerabilities in CRM. In *The Dark Side of CRM: Customers, Relationships and Management*, Nguyen, B., Simkin, L., Canhoto, A.I. (eds). Routledge, New York.

N'Goala, G. (2016). Le marketing dans un monde connecté, un monde de paradoxes… *Décisions Marketing*, 84, 5–18.

N'Goala, G. and Collin-Lachaud, I. (2023). Voyage incongru au Consumer Electronic Show (CES) de Las Vegas. *Décisions Marketing*, 109, 5–12.

Onfray, M. (2001). *Antimanuel de philosophie*. Breal, Paris.

Portes, A. (2018). La transparence numérique : rôle du client et conséquences sur la relation à la marque. PhD Thesis, Université de Montpellier, Montpellier.

Portes, A., Cases, A.S., N'Goala, G. (2017). Vers une définition de la transparence perçue de la relation client sur les canaux digitaux. *Management & Avenir*, 94(4), 105–129.

Portes, A., N'Goala, G., Cases, A.-S. (2020). La transparence numérique : dimensions, antécédents et conséquences sur la qualité des relations clients. *Recherche et Applications en Marketing*, 35(4), 72–98.

Rawlins, B. (2009). Give the emperor a mirror: Toward developing a stakeholder measurement of organizational transparency. *Journal of Public Relations Research*, 21(1), 71–99.

Ruppert, E. (2012). Doing the transparent state: Open government data as performance indicators. In *A World of Indicators: The Production of Knowledge and Justice in an Interconnected World*, Mugler, J. and Park, S.-J. (eds). Cambridge University Press, Cambridge.

Seligman, M.E.P. and Maier, S.F. (1967). Failure to escape traumatic shock. *Journal of Experimental Psychology*, 74, 1–9.

White, R.W. (1959). Motivation reconsidered: The concept of competence. *Psychological Review*, 66(5), 297–333.

5

Involvement

5.1. Introduction: a socially acceptable innovative city at the service of citizen participation

In 2021, 56% of the world's population lived in urban centers (UN-DESA, 2014), and this trend is expected to continue, with this figure expected to reach 66% by 2050 (UN-DESA 2014; UNEP 2018). This rapid urbanization brings new territorial issues with it, as well as a considerable increase in material consumption (Tura and Ojanen 2022). At the same time, only 32.5% of the population in French metropolitan areas say they are interested and involved in what is happening in their city, and only 9.1% actually participate in actions implemented by their local authority (Folcher et al. 2022). For living in a city to still be a positive experience that contributes to the well-being of its inhabitants, the question is then how to identify the channels of communication and public policies that will allow the involvement of city users in a perspective of ecological, social and economic sustainability.

Faced with the demands of growth, performance and competitiveness, cities are responsible for consuming 75% of the world's available natural resources, such as primary energy, raw materials, water and food (UNEP 2012; Tura and Ojanen 2022). It is therefore essential that they ensure effective management of the various negative externalities linked to urbanization (such as resource scarcity, pollution and road congestion). To minimize, correct or even alleviate social, environmental or economic problems, more and more cities are adopting so-called "smart" techniques using relevant and eco-efficient technologies at a societal level, notably by creating more desirable, viable and resilient services and infrastructure networks.

Today, in this period of digital transformation, urbanization is closely linked to technological innovation. As such, the concept of the smart city has been associated

---

Chapter written by Julie JAMMES, Gilles N'GOALA and Pauline FOLCHER.

more widely with innovation due to the diversification of technological components (5G networks, Internet of Things, autonomous vehicles, blockchain, augmented reality, artificial intelligence, etc.) (Beretta 2018). Currently, we can speak of "augmented" cities, thanks, among other things, to the use of digital technologies, and this paradigm offers us important opportunities to reformulate the territories' relationship to the transition to sustainability.

However, the digitization of cities is subject to increasing controversy. In some parts of the world, the concept of a "smart city" is more difficult for society to accept, in that certain uses of data and technological innovation may be seen not only as a means of controlling and monitoring individuals, but also as a factor in dehumanizing users by seeing them more as data collectors and providers than as individuals.

In other words, the socially acceptable innovative city must integrate into its project development and governance the fact that citizens have a threshold of tolerance for changes. This threshold of tolerance refers to a level of psychological saturation, asserting that if local communities perceive and/or experience a deterioration in their quality of life, then they will manifest their frustration through behavior that opposes innovation, thereby leading to the failure of urban development projects (Komsary et al. 2018). Thus, the innovative city must first of all guarantee the need for justice and equity, so that everyone can be made aware of the territory's inherent problems and issues, but also and above all be able to actively participate in territorial governance.

Citizen participation is recognized as a *sine qua non* condition for moving toward the social acceptability of innovation. Indeed, involving civil society in project governance allows us to collectively discuss, define and detect the impacts that innovations will have on the territory. The mode of project governance must therefore, ideally, bring together the diversity of actors present in the territory, particularly if we wish to create civic engagement. However, and as Bourdin's (2020) study attests, project leaders do not always see the usefulness of organizing public consultation sessions. Moreover, they are often satisfied with top-down transmission of information, which leaves little room for citizen expression or debate, thus undermining the very principle of participatory democracy. As a result, from the point of view of local residents, a feeling of frustration may emerge due to lack of consultation from the project leaders or due to lack of information. This governance problem is therefore a challenge that institutional actors must face to establish lasting relationships of trust, but also with the intention of strengthening the democratic basis of the project planning process.

However, the socially acceptable innovative city does not seek to act within deliberative bodies in a utilitarian manner to counter all opposition movements among users, and have an innovation accepted at all costs. On the contrary, an absence of conflict would be synonymous with a symptomatic and apathetic city that would deny citizen participation, thus going against the principle of representative democracy. The socially acceptable innovative city must instead question the foundations of the responses adopted by the public, each of which, at its own level, participates in establishing social innovation. Similarly, it must reflect the factors contributing to making innovation feasible, desirable and equitable.

5.2. The socially acceptable city: toward new registers of legitimacy for innovation

In light of growing evidence that innovation is a source of controversy, opposition and social protest, it becomes necessary to question the attitudes and behaviors adopted by civil society, which is both the subject and the actor of the innovative city.

The study of the practical acceptability of innovation is today largely insufficient to exist on a market, since making a device useful, usable and accessible is a simple prerequisite. Thus, a better understanding of how innovation is socially and psychologically perceived and understood by users is necessary to find new registers of legitimacy.

5.2.1. *The social acceptability of innovation: making the innovative city a socially acceptable living space*

A phrase widely promoted by decision-makers, practitioners and researchers alike, social acceptability of innovation is a polymorphic concept whose contours differ depending on the perspective taken and the social dimension that is understood (at a local, national or even international scale) (Bergeron et al. 2015; Granier 2015).

At the individual level, the social acceptability of technological innovation corresponds to a

> judgment that the individual makes of the feasibility (economic and practical), desirability (benefits/risks), and equity (internal/external) of an innovation based on its impact on society in the broad sense (human, social, environmental, economic, aesthetic, etc.) and which leads them to oppose, support or not react to its dissemination (N'Goala 2021, p. 7).

Thus, there are three dimensions of a judgment of the social acceptability of technological innovation:

– feasibility, which means that innovation must be technically and economically realizable;

– desirability, which insists that innovation must bring a significant level of benefit to society in comparison with the risks it induces;

– equity, which provides for a fair distribution of value among all social entities involved in the implementation of a particular innovation.

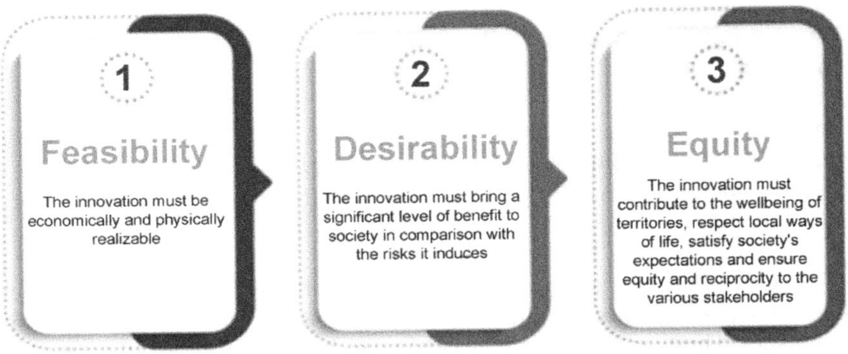

Figure 5.1. *The three dimensions of the social acceptability of technological innovation (adapted from N'Goala 2021)*

Feasibility, desirability and equity are three dimensions that integrate both the technical aspect of innovation and the social aspect of the context through consideration of suprasensible and moral elements. Furthermore, these dimensions support the dynamic, timely and provisional nature of the social acceptability of innovation which responds above all to a logic of social construction (Baba and Raufflet 2014; Gendron 2014).

Indeed, individual judgment is largely shaped by social relationships, shared meanings and value systems that constantly evolve over time and are therefore incompressible to a specific stage in the lifecycle of a technology (Wolfe et al. 2003; Yates et al. 2016; Thomas et al. 2019).

Naturally, this observation emphasizes the "social" component of acceptability which focuses on individuals, their consciousness and their knowledge (Howe et al. 2005). As a result, technology can be integrated through a process of continuous learning, itself regulated by sustained dialog, a principle of equity and social justice, as well as by trilateral trust between promoters, public authorities and citizens (Batellier 2016; Wolsink 2018; Thomas et al. 2019).

5.2.2. Toward a remediation of controversial technology to reach a social compromise

Wolfe et al. (2003) developed the PACT (Public Acceptability of Controversial Technologies) framework, which allows us to distinguish three major dimensions that influence the social acceptability as well as the attractiveness of technology; namely, the constituent dimension composed of the values, motivations and individual and collective strategies of the actors involved in decision-making; the contextual dimension in which the decision-making process takes place; and the technological dimension that centers on the technology involved in comparison with other technological options.

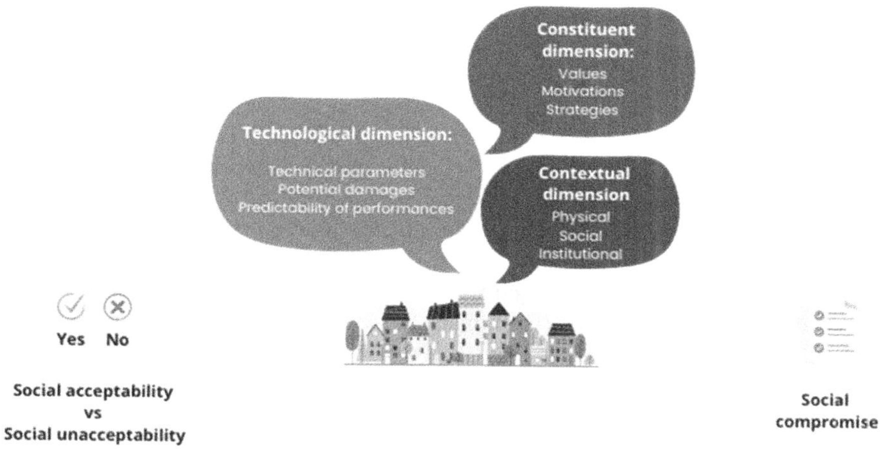

Figure 5.2. *The PACT framework (adapted from Wolfe et al. (2003)). For a color version of this figure, see www.iste.co.uk/ngoala/smartcity.zip*

The constituent dimension integrates values, motivations and strategies. Let us start with a simple postulate: each individual has values which are expressed every day through their interactions and choices (Feather 1995). These values, in part, come to shape basic beliefs, as well as their orientations (intrinsic versus extrinsic). Indeed, when faced with a technology, actors oscillate between the expression of two types of value: one anthropocentric which corresponds to a use value, the other biocentric which refers to a non-use value (Sharp et al. 2012; Ford et al. 2009). Simply put, anthropocentric values instrumentalize the object which is used for the sole purpose of satisfying human desires and needs (Sharp et al. 2012). As a result, any positive consequences related to that object, such as its aesthetic attractiveness or performance, will be deemed more acceptable (Ford et al. 2009). On the contrary, biocentric values make it possible to transcend the instrumental approach to the object which is then considered as an end in itself; for example, even if the object is perceived as not aesthetically attractive, but it respects ethical values and contributes to biodiversity, then it will be judged more acceptable. These two values motivate individuals to learn about a topic or to play an active role in decision-making. They then justify some users engaging in active promotion of (or firm opposition to) this object.

As for the technological dimension, it includes technical parameters, potential damage to human health or environmental health and the predictability of the technology's performance. Strongly linked to stakeholder perceptions, the risks and potential benefits of the technology are inseparable from the social context in which it is deployed (Thomas et al. 2019). Furthermore, several factors influence this perception, including aesthetic and spatial impacts, efficiency, durability, reliability, safety and technological progress (Thomas et al. 2019). These factors themselves depend on social trust, that is, the extent to which the leading figure or the promoter of the technology is perceived as being competent to assess the risks themselves (Bronfman et al. 2012). In this way, actors perceived as being untrustworthy may damage the image of the technology by increasing the perception of risks, which consequently erodes the judgment of social acceptability. In this paradigm, the role of communication is decisive in the sense that it compensates for the lack of information and knowledge, thus allowing stakeholders to establish more impartial assessments of the real extent of the risks and associated benefits while improving social trust in the project leader (Bronfman et al. 2012).

In short, the contextual dimension is marked by three categories: physical, social and institutional attributes. Indeed, any technology evolves within a context, whether physical (topography of a territory, climatic conditions, etc.), social (demographic composition of a community, internal dynamics, etc.) or institutional (the nature of the organizations affected by the technology) (Kim 2016). As a living region, the territory must establish good governance of innovation to understand the elements of context, but also to grasp the social complexities, in order to design a participatory planning structure appropriate to the project (Al-Saidi 2021). Governance of technology within a particular context is critical, as it directly influences public attitudes. In this sense, to establish efficient governance, several factors must be considered: trust in local regulators or in the corporate culture, the possibilities and means offered to citizens to participate in decision-making, transparency of information and the availability of this information (Martínez-Espiñeira et al. 2019).

All three of these dimensions, through their interactions, enrich the dialog oriented toward decision-making in terms of remediation of the controversial technology. In this way, they allow the establishment of collective ground rules to arrive at an acceptable compromise (Wolfe et al. 2003). Social compromise is achieved when there is an informed consensus, that is, the extent to which decision-making includes collective judgment and how this is considered when the technology is implemented (De Coninck et al. 1999). Thus, once the decision-making phase is initiated, it can propel the stakeholders through informed consultation toward a binary and absolute judgment (complete social acceptability versus complete social unacceptability) (Wolfe and Bjornstad 2002).

5.3. The role of communication: a challenge to make the innovative city socially acceptable

Communication is known to directly influence judgments of the social acceptability of innovation, public engagement and participation (Kirkman and Voulvoulis 2017; Saglietto and Charest 2021). As such, project leaders must design communication strategies that account for social considerations to establish lasting relationships of trust.

Furthermore, a project leader's trust capital is even more important when the community is poorly informed about an innovation (Bauwens 2015), since users will assess the legitimacy of the actor solely on the grounds of their history of interactions and their organizational and/or cognitive proximity (Bourdin 2020). Jacques Priol, in Box 5.1, gives concrete examples of the strategies implemented in French metropolises to inform citizens about data collection in public spaces in the hope of their trust and acceptance.

A number of observations and recommendations are derived from previous examples (Nantes, Brest, Angers, Dijon, etc.). First, the innovative city questions ways of engaging the public to encourage interaction and participation to go beyond a top-down approach aiming at passive acceptance of a technology without explaining its foundations and the needs to which it responds.

Without listening, dialog or consultation with the public, the risk for the leaders or promoters of innovation is of incurring wasted expenditure on a project that will not be judged desirable, nor even developed in the right place and at the right time (Kirkman and Voulvoulis 2017).

The case of innovation is particularly delicate, because it implies sociotechnical impacts which are spread over the long term and require a deliberative approach to establish ethical guidelines in decision-making (Cherry et al. 2022).

Citizen participation is thus a key issue in representative democracy and is a minimum condition for ensuring the socially acceptable governance of innovation.

"Smile, you're on camera!" Informing citizens about the deployment of smart technologies in the city

By Jacques Priol[1], president of the CIVITEO firm and president of the Data Publica Observatory

In June 2020, the news shook the world of smart cities: after three years of studies and nearly 50 million Canadian dollars spent, Sidewalk Labs, a sister subsidiary of Google within the Alphabet group, unilaterally withdrew from the smart city project in Toronto, Canada. The Covid-19 pandemic had officially put an end to the model of a city of the future built on an unusual industrial wasteland in the heart of Canada's economic capital. However, the three years of preparation were also marked by many controversies, first local, then national and international. The shadow cast by Google, a digital giant straying into the business of urban operator, sparked protests, first and foremost from citizens joining together in the BlockSidewalk collective. Controversy focused on the capture of data, on the lack of anonymization of these data, on their intended hosting in the United States, thus avoiding Canadian regulations (Priol 2020a). It also focused, despite significant communication efforts, on insufficient citizen involvement in the design of the project.

The central question raised by the failure of the Toronto project is in fact a democratic one. Are the powers of algorithmic calculation and the profusion of data captured in public spaces capable of modeling efficient urban management? And, if so (which remains to be demonstrated), of making citizens accept it? In other words, how can democratic control be guaranteed if management of the major urban functions (organization of travel, management of water, energy and waste, security, etc.) is automated?

This question, obviously, is relevant not only to Toronto. In France, it applies to projects of varying sizes and ambitions. As more and more sensors are deployed in public spaces, whatever the technology used, it is legitimate to wonder: what data are collected? Are they anonymized? Who uses them? How? With which algorithms? With what results? Who can control them? Another question follows, often in an acute manner in France: how can we ensure the social acceptability of these tools (and avoid massive rejection that would give rise to tensions that are quickly publicized and feared by our elected officials)?

Some pioneering communities in France have undertaken work with citizens to inform them, bring them together, involve them and equip them; this is the choice of elected officials. It is interesting to mention some of them.

1 An expert in public data management, since 2016, he has participated in various smart territory projects in France or abroad. Here, he focuses on conditions of information for citizens faced with the proliferation of data capture devices in the public space. He is the author of several works on public data management, including *Ne laissez pas Google gérer nos villes !* (Éditions de l'Aube, 2020), and has published several studies on connected territories, including *Ni surveillance ni consumérisme algorithmique : vers un modèle européen alter-natif pour les villes intelligentes* (Institut français des relations internationales, 2022).

In 2019, the Nantes metropolitan area was the first French local authority to adopt a territorial data charter setting rules (and limits) for the use of data for smart city projects. A citizens' panel was set up, with a clear mandate to define its main principles. In 2021, the Brest metropolitan area went further, by directly entrusting the drafting of its data charter to citizens within the rigorous and demanding framework of a consensus conference.

For their part, the metropolitan areas of Dijon and Angers, which are very advanced in terms of deploying connected city tools, have chosen to establish permanent user committees. The Dijon metropolitan data committee brings together qualified individuals and user associations. The local digital council of Angers, for its part, is directly composed of citizens.

Beyond these participatory bodies, there is also the question of informing citizens in the public space. The exhibition "Smile, you're on camera!", which was presented in Nantes and Bordeaux, is an initiative of private companies keen to transparently explain the tools deployed in public spaces. Obligations to inform the public are in fact very weak, with the notable exception of video surveillance cameras. However, the sensors used are increasingly numerous, often very discreet, and their uses very varied. Also, some are not without risks (Priol 2020b)!

Involving citizens in defining the rules of the game in advance (in Nantes, Brest or Angers) or involving them in the governance of a project that is already widely deployed (in Dijon): both options have their advantages and disadvantages. These initiatives are also embedded in a regulatory context that is still imperfect. The obligations for algorithmic transparency enforceable against public actors are not well known by local authorities. They are also very incomplete and very imperfect (Bourcier and De Filippi 2018). Above all, they are very far removed from the concerns of residents who move through a public space increasingly equipped with data collection technologies.

To conclude, it is interesting to return to Toronto. Faced with criticism and controversy, Google's teams had prepared a system to inform residents in the form of urban signage indicating all of the sensors and providing precise information, via a QR code, on the purposes and methods of using the data. A few years after the Toronto project was abandoned, these tools are being reused by pioneering stakeholders and territories in the form of an open-source standard, notably tested in 2022 in Boston and Washington, D.C. in the United States, Innisfil in Canada and Angers in France. The initiative for this activation came from a Canadian non-profit organization (Helpful Places, n.d.). This standardized signage, designed by designers of many nationalities, has now been translated into French. It is available under an open license. The "DTPR" standard (see Chapter 4) is thus undoubtedly, in France and Europe, the most advanced tool available to date for territories wishing to provide the most complete transparency on the technologies deployed in the city. It is a useful complement to consultation and participation processes limited by design to a restricted number of users. Without transparency, the smart city in fact risks quickly becoming automatic and anti-democratic.

Box 5.1. *"Smile, you're on camera!" Informing citizens about the deployment of smart technologies in cities by Jacques Priol*

Representative democracy relies on decision-making that responds to principles of justice and equity to create legitimacy for the decision-making process. In other words, the socially acceptable innovative city must guarantee conditions conducive to the development of community well-being, which necessarily presupposes a good governance structure and a certain level of citizen representativeness.

Furthermore, it is through these guarantees that the link between citizen participation and decision-making will be founded on a socially acceptable basis, and which, consequently, will enrich democratic deliberations (Feurtey et al. 2010; Granier 2015).

Indeed, a decision that is intended to be fair will be judged to be more effective, and will therefore be better accepted by all, because a fair procedure tends to produce fair results (Field et al. 2012; Batellier 2016).

However, citizen participation requires a high degree of engagement to lead to a fruitful consultation phase, or even a co-decision phase punctuated by a full uptake of essential knowledge so that everyone can take a position on a decision which has societal repercussions (Yates and Caron 2013). Through this perspective, elected officials, local authorities and experts must act together to meet the challenge of social acceptability, in particular by implementing three phases: research and preparation, persuasive communication and awareness-raising, and taxation and policy (Singer 2013; Marciano et al. 2014).

The research and preparation phase aims to provide information to citizens so that they can have knowledge about the project and thus judge its level of expertise and reliability (Bourdin et al. 2019). Indeed, a lack of information and knowledge is detrimental to social acceptability (Reig et al. 2019), as it increases perceptions of risk for the technology (Karimi and Toikka 2018) and minimizes perceptions of its benefits for society (Kirkman and Voulvoulis 2017). As a result, transparent and sustained provision of information over time should help alleviate the feeling of mistrust by breaking down potential myths related to the technological artifact (Reig et al. 2019; Saglietto and Charest 2021).

The persuasive communication and awareness phase relates to the effectiveness and scope of the communication campaign that is conducted. This must be targeted and contain short, simple and convincing statements, emphasizing the needs to which the project responds while indicating the impact it will have on users' quality of life, that is, the consequences and results expected during its implementation (Ford et al. 2009; Kirkman and Voulvoulis 2017). However, stakeholders on the ground point out that citizens do not require complete information to take a position. Some evaluate a previously unknown technology in the form of a "pseudo-opinion without information" based on their intuitions and feelings (Huybens 2011). In this case, trust in the promoters of technology will be an alternative to detailed information (Gordon et al. 2014).

Finally, the taxation and policy phase embodies the institutional process, that is, the procedure for dialog between the actors. The objective is to arrive at a social compromise that respects everyone's interests (Bergeron et al. 2015). Politics, being in itself a deliberative and participatory process, can also be deployed in the digital space, thus granting citizens the possibility of extending their rights and making their voices heard. This active and digital participation, enabled by a responsible use of technology, allows actors to benefit from a digital citizenship in addition to their so-called everyday citizenship.

5.4. Effectively raising awareness among users about innovation

There are three dimensions that allow a practice to build legitimacy: legal legitimacy, social acceptability and ecological credibility (McLain et al. 2018). However, social acceptability today still lacks a legislative and regulatory framework (Batellier 2016), despite its undeniable interest in providing decision-makers and practitioners with good practices to make innovation socially acceptable.

As a result, the grounding and contours of social acceptability are still to be defined. While it is interesting to note the various constituent factors affiliated with it, nevertheless there is still a lack of clarity around how to define it and understand it operationally. Despite this, existing research on the subject still allows us to identify five good practices for effectively raising users' awareness of innovation.

First, to raise awareness among civil society, it may be useful to mobilize experts who will strive to present detailed information with expertise and a certain amount of credibility. In this sense, when a large-scale project is being considered in a territory, organizing discussion groups accessible to all for communicating about evidence (key figures, graphs and images) can be a great support for considerably increasing public confidence (Shackley et al. 2009; Kirkman and Voulvoulis 2017).

Second, it may be relevant to focus information on the potential benefits and costs of the technology, translating the impact of its use on a personal, social and environmental level (Thomas et al. 2019). The goal is to produce effective, clear and impartial communication.

Furthermore, if the technology relates to considerations of environmental protection, communications should use elements of language that highlight this type of value (Wagner et al. 1998), especially to create an echo among that section of the public, which has mainly biocentric values.

Third, decision-makers and promoters of innovation must never communicate information in an isolated, one-way manner. An oversimplified message can increase the polarization of responses among users, who may then express either total agreement or total opposition. Similarly, polarization of public responses is largely influenced by the perceived reliability of the information source, by the tone used in the message, but also by unexpected exposure to contradictory messages on the same subject. Therefore, to raise awareness among civil society, a communication strategy must be carefully crafted and considered at length. In this context, it is particularly relevant to use different messages and different media which strive to begin a dialog to reach as many people as possible.

However, there is a risk of encountering the Dunning–Kruger effect (Kruger and Dunning 1999), meaning that some people may have a reduced willingness to process new information, believing that they already have sufficient background to take a position. Thus, paradoxically, improving knowledge about an innovation may require convincing users that there are still many things to learn (Mylek and Schirmer 2019).

Fourth, certain actors, playing the role of facilitator and/or intermediary, are considered more legitimate in providing information. For example, local authorities as intermediary actors can initiate a dynamic of territorial proximity by arbitrating the divergent interests of developers and users (Bourdin et al. 2019).

Similarly, some public–private partnerships between those who hold knowledge and those who hold resources can contribute to improving the sharing of technical knowledge with agents present on the ground, who are the main embodiments of the role of prescribers of opinion (Okello et al. 2021).

Finally, it becomes necessary for decision-makers to anticipate, as far as possible, the concerns of users and societal conditions, especially for innovations which have a presumed impact on health and/or the environment (Thomas et al. 2019). To do this, the use of academic research, especially in the human sciences, may be interesting. The use of *design fiction* techniques, increasingly used by companies and public institutions, would help define the vision that users may have of the consequences of innovation for their environment, their city or quite simply their everyday life.

Given the context and the various issues linked to the development of smart cities, we can conclude this chapter by returning to the conditions necessary for the social acceptability of these projects. Development of technologies in the heart of cities must be perceived as feasible, desirable, and equitable by stakeholders, but this is not enough. The legitimacy of project leaders, the trust placed in them by stakeholders, but also the space left open for consultation and citizen participation seem essential.

Finally, the communication deployed to enable this user involvement must not only be easily accessible and available, but it must also guarantee transparency and mutual listening in exchanges.

5.5. References

Al-Saidi, M. (2021). From acceptance snapshots to the social acceptability process: Structuring knowledge on attitudes towards water reuse. *Frontiers in Environmental Science*, 9, 50.

Baba, S. and Raufflet, E. (2014). Les relations entreprises-communautés : les leçons de l'expérience d'Hydro-Québec et des Cris. *Gestion*, 39(2), 104–111.

Batellier, P. (2016). Acceptabilité sociale des grands projets à fort impact socioenvironnemental au Québec : définitions et postulats. *VertigO – La revue électronique en sciences de l'environnement*, 16(1). doi: 10.4000/vertigo.16920.

Bauwens, T. (2015). Propriété coopérative et acceptabilité sociale de l'éolien terrestre. *Reflets et perspectives de la vie économique*, 54(1), 59–70.

Beretta, I. (2018). The social effects of eco-innovations in Italian smart cities. *Cities*, 72, 115–121.

Bergeron, K. M., Jébrak, M., Yates, S., Séguin, C., Lehmann, V., Le Meur, P. Y., Angers, P., Durand, S., Gendron, C. (2015). Mesurer l'acceptabilité sociale d'un projet minier : essai de modélisation du risque social en contexte québécois. *VertigO – La revue électronique en sciences de l'environnement*, 15(3). doi: 10.4000/vertigo.16737.

Bourcier, D. and De Filippi, P. (2018). La transparence des algorithmes publics. Observatoire Data Publica. *Les cahiers de l'Observatoire*, 3.

Bourdin, S. (2020). Le NIMBY ne suffit plus ! Étude de l'acceptabilité sociale des projets de méthanisation. *L'Espace Politique*, 38. doi: 10.4000/espacepolitique.6619.

Bourdin, S., Nadou, F., Raulin, F. (2019). Les collectivités locales comme acteurs intermédiaires de la territorialisation de la transition énergétique : l'exemple de la méthanisation. *Géographie, économie, société*, 21(4), 273–293.

Bronfman, N.C., Jiménez, R.B., Arévalo, P.C., Cifuentes, L.A. (2012). Understanding social acceptance of electricity generation sources. *Energy Policy*, 46, 246–252.

Cherry, C., Thomas, G., Groves, C., Roberts, E., Shirani, F., Henwood, K., Pidgeon, N. (2022). A personas-based approach to deliberating local decarbonisation scenarios: Findings and methodological insights. *Energy Research & Social Science*, 87, 102455.

De Coninck, P., Séguin, M., Chornet, E., Laramée, L., Twizeyemariya, A., Abatzoglou, N., Racine, L. (1999). Citizen involvement in waste management: An application of the STOPER model via an informed consensus approach. *Environmental Management*, 23(1), 87.

Feather, N.T. (1995). Values, valences, and choice: The influences of values on the perceived attractiveness and choice of alternatives. *Journal of Personality and Social Psychology*, 68(6), 1135.

Feurtey, É., Saucier, C., Côté, G., Jean, B. (2010). L'acceptabilité sociale de l'énergie éolienne au Québec : quel rôle pour les élus municipaux ? *Revue organisations & territoires*, 19(3), 5–16.

Field, C.B., Barros, V., Stocker, T.F., Dahe, Q. (2012). *Managing the Risks of Extreme Events and Disasters to Advance Climate Change Adaptation: Special Report of the Intergovernmental Panel on Climate Change*. Cambridge University Press, Cambridge.

Folcher, P., Busca, L., Münzel, A., Pellequier, V. (2022). *Baromètre de l'intelligence urbaine vécue : regards des habitants sur leur métropole*, 1st edition. Chaire Cit.Us, Université de Montpellier – MOMA.

Ford, R.M., Williams, K.J., Bishop, I.D., Hickey, J.E. (2009). Effects of information on the social acceptability of alternatives to clearfelling in Australian wet eucalypt forests. *Environmental Management*, 44(6), 1149–1162.

Gendron, C. (2014). Penser l'acceptabilité sociale : au-delà de l'intérêt, les valeurs. Communiquer. *Revue de communication sociale et publique*, 11, 117–129.

Gordon, R., Brunson, M.W., Shindler, B. (2014). Acceptance, acceptability, and trust for sagebrush restoration options in the Great Basin: A longitudinal perspective. *Rangeland Ecology & Management*, 67(5), 573–583.

Granier, B. (2015). L'expérimentation sociotechnique fondée sur les sciences comportementales : un instrument au service de la production de l'acceptabilité sociale ? *VertigO – La revue électronique en sciences de l'environnement*, 15(3). doi: 10.4000/vertigo.16695.

Helpful Places (n.d.). Helpful places: Embed transparency into your smart cities [Online]. Available at: https://helpfulplaces.com [Accessed April 3, 2025].

Howe, G.T., Shindler, B., Cashore, B., Hansen, E., Lach, D., Armstrong, W. (2005). Public influences on plantation forestry. *Journal of Forestry*, 103(2), 90–94.

Huybens, N. (2011). Comprendre les aspects éthiques et symboliques de la controverse socioenvironnementale sur la forêt boréale du Québec. *VertigO – La revue électronique en sciences de l'environnement*, 11(2). doi: 10.4000/vertigo.11119.

Karimi, F. and Toikka, A. (2018). General public reactions to carbon capture and storage: Does culture matter? *International Journal of Greenhouse Gas Control*, 70, 193–201.

Kim, E.J. (2016). Phytoremediation for lightly toxic sites: Hazard perception and acceptance of remediation alternatives. *Human and Ecological Risk Assessment: An International Journal*, 22(4), 1078–1090.

Kirkman, R. and Voulvoulis, N. (2017). The role of public communication in decision making for waste management infrastructure. *Journal of Environmental Management*, 203, 640–647.

Komsary, K.C., Tarigan, W.P., Wiyana, T. (2018). Limits of acceptable change as tool for tourism development sustainability in Pangandaran West Java. *IOP Conference Series: Earth and Environmental Science*, 126(1), 012129.

Kruger, J. and Dunning, D. (1999). Unskilled and unaware of it: How difficulties in recognizing one's own incompetence lead to inflated self-assessments. *Journal of Personality and Social Psychology*, 77(6), 1121.

Marciano, J.A., Lilieholm, R.J., Teisl, M.F., Leahy, J.E., Neupane, B. (2014). Factors affecting public support for forest-based biorefineries: A comparison of mill towns and the general public in Maine, USA. *Energy Policy*, 75, 301–311.

Martínez-Espiñeira, R., Garcia-Valinas, M.A., Matesanz, D. (2019). Public attitudes towards hydraulic fracturing in Western newfoundland. *Energy Economics*, 84, 104492.

McLain, R., Lawry, S., Ojanen, M. (2018). Fisheries' property regimes and environmental outcomes: A realist synthesis review. *World Development*, 102, 213–227.

Mylek, M.R. and Schirmer, J. (2019). Thinking about fuel management: The potential of integrative complexity theory to inform design of communication about fuel management used to reduce wildfire risk. *Society & Natural Resources*, 32(9), 983–1002.

N'Goala, G. (2021). L'acceptabilité sociale d'une innovation technologique : un essai de conceptualisation et de modélisation ? *Congrès de l'association française de marketing*, Angers.

Okello, A.O., Nzuma, J.M., Otieno, D.J., Kidoido, M., Tanga, C.M. (2021). Farmers' perceptions of commercial insect-based feed for sustainable livestock production in Kenya. *Sustainability*, 13(10), 5359.

Priol, J. (2020a). *Ne laissez pas Google gérer nos villes*. Éditions de l'Aube, La Tour d'Aigues.

Priol, J. (2020b). "Souriez, vous êtes captés !". Livret de l'exposition [Online]. Available at: https://souriezvousetescaptes.fr.

Raufflet, E. (2014). De l'acceptabilité sociale au développement local résilient. *VertigO – La revue électronique en sciences de l'environnement*, 14(2). doi: 10.4000/vertigo.15139.

Reig, L., Escobar, C., Carrassón, M., Constenla, M., Gil, J.M., Padrós, F., Flos, R. (2019). Aquaculture perceptions in the Barcelona metropolitan area from fish and seafood wholesalers, fishmongers, and consumers. *Aquaculture*, 510, 256–266.

Saglietto, L. and Charest, F. (2021). Technologies innovantes et communication d'acceptabilité. *Communication & Organisation*, 59, 263–277.

Shackley, S., Reiner, D., Upham, P., de Coninck, H., Sigurthorsson, G., Anderson, J. (2009). The acceptability of CO_2 capture and storage (CCS) in Europe: An assessment of the key determining factors: Part 2. The social acceptability of CCS and the wider impacts and repercussions of its implementation. *International Journal of Greenhouse Gas Control*, 3(3), 344–356.

Sharp, E.A., Spooner, P.G., Millar, J., Briggs, S.V. (2012). Can't see the grass for the trees? Community values and perceptions of tree and shrub encroachment in south-eastern Australia. *Landscape and Urban Planning*, 104(2), 260–269.

Singer, C. (2013). Stakeholder attitudes toward forest-residual based biofuels in Washington State. PhD Thesis, Clemson University, Clemson.

Thomas, G., Demski, C., Pidgeon, N. (2019). Deliberating the social acceptability of energy storage in the UK. *Energy Policy*, 133, 110908.

Tura, N. and Ojanen, V. (2022). Sustainability-oriented innovations in smart cities: A systematic review and emerging themes. *Cities*, 126(1), 103716.

UN-DESA (2014). World urbanization prospects: The 2014 revision, highlights. Report, United Nations Department of Economic and Social Affairs.

UNEP (2012). Sustainable, resource efficient cities – Making it happen! Report, United Nations Environment Programme.

UNEP (2018). The weight of cities – Resource requirements of future urbanization. Report, United Nations Environment Programme.

Wagner, R.G., Flynn, J., Mertz, C.K., Slovic, P., Gregory, R. (1998). Acceptable practices in Ontario's forests: Differences between the public and forestry professionals. *New Forests*, 16(2), 139–154.

Wolfe, A.K. and Bjornstad, D.J. (2002). Why would anyone object? An exploration of social aspects of phytoremediation acceptability. *Critical Reviews in Plant Sciences*, 21(5), 429–438.

Wolfe, A.K., Bjornstad, D.J., Kerchner, N.D. (2003). Making decisions about hazardous waste remediation when even considering a remediation technology is controversial. *Environ. Sci. Technol.*, 37(8), 1485–1492.

Wolsink, M. (2018). Social acceptance revisited: Gaps, questionable trends, and an auspicious perspective. *Energy Research & Social Science*, 46, 287–295.

Yates, S. (2013). La communication comme vecteur de l'acceptabilité sociale des grands projets. *Journal of Professional Communication*, 2(2). doi: 10.15173/jpc.v2i2.128.

Yates, S., Bergeron, K. M., Jébrak, M., Angers, P., Lehmann, V., Séguin, C., Gendron, C. (2016). Indice du risque social : un outil pour mieux saisir les enjeux, risques et opportunités des projets miniers. *Éthique publique – Revue internationale d'éthique sociétale et gouvernementale*, 18(1), 1–24. doi: 10.4000/ethiquepublique.2417.

6

Informing

6.1. Introduction: fighting the spread of fake news in the smart city

"A humped shape is rising out of the pit. […] What's that?!": on October 30, 1938, the American radio station CBS brought us live coverage of an extraterrestrial landing. Orson Welles' hoax, through his radio adaptation of *The War of the Worlds*, the famous book by H.G. Wells, is one example of fake news that has certainly made one of the biggest impressions to date. Since then, there have been successive disinformation campaigns, sometimes with harmful effects on citizens, businesses, the media, governments and society as a whole. In the contemporary digital context, this false information circulates at high speed on the web and is particularly relayed through the various forms of social media.

While it is difficult to reach a consensus among researchers and professionals on the definition of the phenomenon represented by fake news, we consider it as follows: it involves "intentionally false, realistic and fabricated stories which can potentially be verified" (Visentin et al. 2019, p. 99). Here, it is interesting to note that this misleading information results from a malicious intention, that of deceiving its recipients. Also, these misleading articles imitate those published by recognized conventional media in order to benefit from the same credibility. Furthermore, since the main objective is to create widespread doubt to confuse the greatest number of individuals, fake news often mixes truth and falsehood to blur the line between reality and fiction.

In the smart city, communication and information technologies play a key role in continuous improvement of its operations (Batty et al. 2012). The spread of false

---

Chapter written by Ambre GAMBIN.

For a color version of all figures in this chapter, see www.iste.co.uk/ngoala/smartcity.zip.

information therefore poses a definite threat to the development of these innovative agglomerations. Misleading information may also act as a real barrier to innovation by discrediting innovative projects. For example, electric vehicles are subject to vigorous disinformation campaigns: in reality, it is said, they are more polluting than thermal vehicles, less safe, have non-recyclable batteries and a short range. This fake news is intended to discourage potential consumers from considering purchasing this product. We can also mention another conspiracy theory that targets certain smart cities. "15-minute cities" anticipate all essential services being made accessible to citizens on foot or by bicycle. However, some conspiracy theorists claim that this is in fact a ploy by governments to restrict the movements of individuals, increase surveillance and limit individual freedoms. In the smart city, it therefore seems necessary to mobilize technological and human resources to effectively combat disinformation.

6.2. A technological environment facilitating the spread of misleading information

6.2.1. *The central role of social media in disseminating fake news online*

Today, fake news is mostly propagated through social networks. In smart cities, these platforms represent a major source of information for citizens. According to Gottfried and Shearer (2016), 62% of adults get their news daily from these digital media. By bringing together millions of active users across the world, these networks contribute to rapid and large-scale dissemination of information of all kinds. By 2023, 59.4% of the world's population was registered on at least one of these platforms, or a total of 4.76 billion users (We Are Social and Hootsuite 2023). To interact with their loved ones, social media users are regularly encouraged to share content online. Studies estimate that 23% of users have already contributed to spreading misleading information, intentionally or not. However, on the web, fake news spreads faster than real news (Vosoughi et al. 2018).

This phenomenon may have several explanations inherent to the specific functioning of these platforms. First, unlike traditional media, social media has few (if any) barriers to entry. Any individual is thus invited to register, then to become an active member interacting with the information present on these sites by expressing their interest in it through likes, comments or shares. Each user can also be part of the creation of information. Next, these platforms offer a varied set of posts (messages, articles, audiovisual content, etc.) to users based on various criteria. So, when they log into their account, they encounter content preselected by the site's algorithm. This content consists of articles with which the user's follows/followers have already interacted, suggestions based on their previous activity, but also posts that are exclusively advertising. Individuals therefore do not always control the information to which they are exposed, making its true source hard to identify. Furthermore, these platforms are commercial entities. As a result, they constantly seek to maximize user satisfaction by offering targeted and personalized content. This is based on users' past preferences and behaviors, in line with their opinions. This leads to the creation of "echo chambers", or "information bubbles", which reinforce individuals in their beliefs. It may sometimes happen that users encounter conflicting information, generating cognitive dissonance. In this case, most individuals tend to reject this uncomfortable content, or even consider it false. In this sense, on social networks, members interact with and publish information that is consistent with their opinions and beliefs, which they consider credible and reliable: this is confirmation bias. In today's divided context, social media finds itself polarized, as are its users. This bias unconsciously encourages users to accept only confirmatory information, even if it contradicts established facts.

While social media, by its nature, contributes greatly to the spread of fake news online, its users are equally responsible. These can be categorized into two distinct groups: individuals with malicious intent, on the one hand, and those with benevolent intent, on the other hand. The first category includes members of these platforms who deliberately share content they know to be false, mainly for political or ideological purposes. Some may be paid to spread particular information or to target a specific demographic. The second category includes users who spread false information unintentionally, relying on it and considering it truthful. These individuals generally share content when it contains information judged to be of high quality. This perception of quality plays a crucial role in the dissemination of information online, especially when it is coupled with a high level of trust in its source. Additionally, social media users are influenced by the behaviors of their peers and the need to conform to them. They devote little time or cognitive effort to reading and assimilating virtual content. Interactions (sharing, liking, comments) from other members of the platform play a significant role in their attitudes toward false information. Social impact and collective opinion may then promote the dissemination of fake news by users with initially benevolent intentions. Therefore, the sharing of misleading information may not only be initiated by a desire to accomplish a specific political or ideological agenda, but also by a need for social acceptance and homophily intrinsic to the virtual community of belonging or aspiration (Domenico et al. 2021). Some studies have also emphasized the harmful impact of social networks on individual behavior. These platforms may breed vindictive, aggressive, self-centered and narcissistic behavior. They may affect an individual's cognitive, emotional and social aspects, and even alter their mental health. This dark side of social media may, in the long term, significantly damage the well-being and freedom of communities and individuals (Talwar et al. 2019).

Social networks, as well as their users, play a decisive role in the spread of fake news online. In recent years, certain technologies have appeared on these platforms that contribute to the creation of massive disinformation campaigns.

6.2.2. *Involvement of automated technologies in the distortion of reality*

While active members of various social media outlets share a good amount of false information, the sources behind their publication are not necessarily human. In fact, the producers of fake news depend mostly on bots to make their content go viral online.

Social media bots are described as registered and active accounts on social networks, controlled entirely (or in part) by computer algorithms (Yang et al. 2019). They can generate content automatically, but also interact with other human users, often imitating them or posing as one of them. Some are harmless, even useful, in certain situations where they save manual labor without interfering with human exchanges. However, they can also surreptitiously manipulate public discourse and influence the attitudes and behaviors of real users. While more traditional nefarious entities primarily target hardware and software vulnerabilities, bots exploit human vulnerabilities on social media. Multiple bot accounts can be controlled by a single malicious entity and quickly generate content, amplifying alternative facts or misinformation. They can deceive humans, but also algorithms based on engagement and interaction, by giving the impression that an opinion or a person is growing in popularity on the platform.

Social bots are continually evolving in response to the measures deployed by platforms to eliminate them, so that they appear ever more sophisticated and humanlike. Often, social bots come together in networks. These groups are called botnets and act in a coordinated manner, posting the same content repeatedly with the aim of making it go viral. This type of behavior is generally difficult to detect, as the accounts concerned appear authentic when analyzed individually. It becomes noticeable only if we can distinguish unusual and organized activity. By artificially mimicking actions such as liking, sharing and commenting on online content, bots are known to accelerate the spread of fake news. Bots, as well as botnets, are active in different areas, affecting individuals' behaviors, attitudes, opinions and choices. They can steer the debate around major social and societal issues at the heart of current events and often involving political or public health problems.

Bots and botnets represent a clear danger regarding the deliberate and automated propagation of fake news on a large scale. However, a new threat using artificial intelligence (AI) has recently emerged and is the source of many virtual frauds: deepfakes. These are defined as digitally manipulated synthetic media content (videos, images, sound clips, etc.) in which individuals are shown saying or doing something that never happened in the real world (Westerlund 2019). Deepfakes appear to be true to reality and enjoy strong credibility due to the difficulty of distinguishing them from authentic content. This increasingly accessible technology shows a worrying potential for disruption. Some synthetic media can generate realistic and convincing faces of humans who have never walked the Earth (see Figure 6.1).

Figure 6.1. *Portraits generated by AI*[1]

On the Internet, deepfakes contribute greatly to the phenomenon of fake news online and on social networks by creating doubt among users, which in turn affects their relationship with information and the media. In fact, this process leads individuals into error. Furthermore, even when they are not fooled, they may experience future uncertainties about their ability to distinguish truth from falsehood in other media content. A study by Vaccari and Chadwick (2020) shows that not only do deepfakes mislead individuals, but they reduce their trust in information circulating on social networks. In the long term, these effects may have disastrous repercussions for online civility and cause certain problematic behaviors. Citizens are less likely to cooperate in a context of mistrust, a fortiori in the currently highly polarized political and media environment. This would have the effect of drastically reducing the adoption of collaborative and responsible attitudes toward other users. Some individuals may seek to completely avoid potential exposure to news and current events. In this situation, we might imagine that public debate would become more difficult and might be subject to sabotage, endangering democracy.

1 See: thispersondoesnotexist.com.

6.3. Technological innovation in the service of detecting false information

6.3.1. *Identifying fake news using artificial intelligence*

AI and various spin-off technologies can pose a threat to the spread of fake news online at the heart of the smart city. However, in recent years, the use of AI as a detector of false information on social networks has evoked keen interest in scientific and professional communities. Indeed, whether it involves AI, machine learning (ML) or deep learning (DL), these innovations are increasingly used to create predictive systems capable of spotting fake news across social media. Approaches to detecting misleading information have recently been classified into two categories: content-based models and social context-based models. Both processes usually integrate ML technologies.

Methods belonging to the first class (content-based) focus on the information content: the body of the text, the title and certain additional metadata (when available). They represent the approach traditionally adopted, as they appear particularly suited to traditional media and apply more generally to all cases in which no social information is present. Historically, they were used in the context of detecting spam in emails and web pages, and have recently been applied to the detection of fake news. The main difficulty in implementing content-based methods lies in the fact that fake news, as we contemplate it, is based on a deliberate and malicious intent to deceive those exposed to it. It mimics the format and uses the codes of traditional information, clouding the reality of the facts. Furthermore, while these approaches sometimes make it possible to effectively distinguish between highly polarized articles and those intended for the general public, they do not in themselves solve the overall problem of detecting misleading information online. To date, their possibilities of use remain rather limited.

Methods specific to the second category (social context-based) focus on data relating to social signals, such as users' engagement or interaction with a given content on social networks (liking a post on Facebook, retweeting information on X (formerly Twitter), commenting on a video on Instagram, etc.). They constitute a more recent strategy and have a strong potential for detecting fake news, as they assess the credibility of content circulating through social media based on its social context. Other techniques related to these use mapping of the information diffusion model. Indeed, the democratization of social networks has given a new dynamic to information sharing. Users have a natural tendency to group together in communities and to develop within "echo chambers". In this case, some ML algorithms can, in addition to the social context of the post, also consider the specific characteristics of the user interacting with the message to better categorize it. However, social context-based approaches are inherently constrained by the availability of social signals. They cannot be applied in the absence of these characteristics and are less effective when social interactions are limited or few in number. This may happen frequently, especially when content has just been published, thus compromising the early detection of fake news on social networks.

In addition to detecting fake news on social networks, AI also helps identify the bots that create and relay it. Various characteristics are used to recognize these accounts and distinguish them from those belonging to real users, such as account metadata, metadata of friends, followers or subscriptions, or the structure of the sharing network. After being extracted and pre-processed, these data are fed into ML models, which are then used to test user accounts across various platforms. In 2014, researchers at Indiana University developed a service to detect bots on X: the Botometer. Using a ML approach, this algorithm is trained to extract user account-specific features from data provided by X's Application Programming Interface (API). It then produces a bot score to classify the analyzed accounts: the higher the score, the greater the probability that the account is controlled entirely or in part by software.

6.3.2. *The importance of content moderation on social media*

AI and its different modes of learning, whether ML or DL, effectively make it possible to detect content, as well as suspicious accounts on social networks. As these platforms are part of the tech industry, for several years, they have adopted approaches aimed at automatic identification of misleading and problematic information using their respective algorithms. This is then downgraded or even deleted so that as few users as possible are likely to be exposed to it: this is content moderation. According to Grimmelmann (2015), content moderation can be defined as a governance mechanism that structures participation in a community in order to facilitate cooperation and prevent potential abuse. Today, many sites, such as Wikipedia, Twitch and Reddit, use moderation systems that integrate automated tools. However, this approach confronts two important limitations. First, truth is not a clearly defined concept. Even professionals have disagreements when it comes to fact-checking information. It is therefore complicated to decide which content and features should be included in these automated systems, since techniques using AI can make mistakes and perform censorship that later turns out to be unjustified. Also, in the virtual environment that we know, the content of disinformation campaigns tends to evolve at an accelerated pace. Consequently, the tools and criteria currently used and considered effective for identifying fake news may quickly become obsolete. This may occur abruptly when certain major social or societal crises suddenly arise, such as the Covid-19 pandemic.

Most social media, especially the most popular ones, already had algorithms allowing them to moderate the content present on these platforms. However, since the recent health crisis, most of them strengthened existing measures with the aim of containing the spread of fake news. For example, Meta announced in late 2020 that it was committed to removing false claims about the coronavirus vaccine, already debunked by public health experts, from Facebook and Instagram. The Meta group has also upgraded its automated moderation system. In fact, the company uses AI for both its Facebook and Instagram platforms. Thanks to its learning processes, AI is at the real heart of the content review process, being able not only to detect, but also to remove those violating the rules established by the platform. The algorithm is programmed so that problematic information is removed, even before it is reported by one or more users. Meta uses an AI approach based on a ML model capable of recognizing and analyzing the intrinsic elements of different types of content (image, video, text, etc.). Depending on the case, the model can reduce the diffusion, or even remove the content directly, or transmit it to a team of reviewers who will perform in-depth analysis. In this latter situation, this group will make the final decision, at the same time allowing the automated system to learn from it so it can improve in anticipation of future interventions. In this sense, Meta has also implemented moderation measures to curb the spread of fake news on its other entity, WhatsApp. In fact, the application now features a new restriction which prevents its users from sending messages already forwarded multiple times (more than five times) to more than one chat group. This provision is intended to limit the sharing of spam or messages containing false information.

As for X, the site has developed a "quality filter" that attempts to predict whether content posted and shared is low-quality information, spam messages, or fully automated posts. The platform has taken a firm stance on freedom of expression and shows hesitation in strictly moderating its content. This filter was designed to reduce the visibility of the content concerned, not to remove it. If such content is detected, user interactions with the post will be limited, or even made impossible, to reduce its chances of going viral. For its part, TikTok has developed a technology similar to that of Meta. Videos uploaded to the app are evaluated by AI to detect content violating its guidelines. This system considers several features such as images, titles, descriptions, keywords and audio. If a violation of the rules is found, the algorithm can delete the content in the case of a high and obvious threat, or forward it to designated security teams for further review if the breach appears less urgent or serious. Finally, YouTube has deployed an automated moderation strategy based on four main axes called the "4 Rs": removing content infringing on community rules, reducing recommendations of content that is on the verge of infringement, relaying reliable sources and rewarding trusted creators. For each of these measures, the platform uses AI based on automatic models (ML).

While social networks are working to develop more precise, clear and effective algorithms and automated moderation systems, most experts warn that these programs are vulnerable and may be subject to errors and numerous biases. A Pew Research Center study found that 69% of respondents would rather have content reviewed for a longer time but with more accuracy, rather than have it reviewed quickly but potentially have useful and accurate information removed in error. According to them, priority must be given to accuracy, even if it means some potentially misleading information remaining longer on platforms. Furthermore, most participants felt that decisions regarding misinformation on these networks should be taken in conjunction with manual participation. They express a favorable opinion on the use of hybrid approaches combining AI and human teams. Only 6% of citizens believe that these judgments should be carried out exclusively by automated computer programs (Rainie et al. 2022).

6.4. Protecting yourself from fake news in the smart city: the need for human resources

6.4.1. *Setting the record straight with fact-checking*

While the intervention of automated technologies, such as AI, represents a promising advance in the fight against fake news online, some gaps remain regarding these innovative approaches. To circumvent them, many actors ranging from social networks to traditional media have implemented manual techniques dedicated to slowing the spread of false information. The most widely used method today is fact-checking. This is one of the so-called reactive approaches (debunking). They therefore intervene after previous exposure to misleading information. The Cambridge Dictionary defines fact-checking as follows: it is “The process of checking that all the facts in a piece of writing, a news article, a speech, etc., are correct”. This practice is currently widely used. For example, Facebook has implemented its own fact-checking strategy since 2016. Since then, other platforms, such as Instagram and X, have done likewise. Many news sites, such as Le Monde, France Info and AFP, have also created a section dedicated to verifying facts making the news. Other actors exclusively dedicated to fact-checking have also appeared online, such as Snopes or PolitiFact.

Fact-checking can reduce the impact of misinformation on social media in two ways. On the one hand, it informs users of the inaccuracy of a piece of information. This has the effect of restricting incorrect beliefs and intentions to share. On the other hand, platforms can, in addition to their algorithm, control the quality of the information that users will encounter. To increase the effectiveness of fact-checking, research recommends certain practices. First, it is important to cite verifiable facts and provide an alternative causal explanation. Next, it is necessary to use simple and informative language, possibly accompanied by explanatory illustrations. Finally, the correction must not become a threat to the interlocutor's ideology, so as not to endanger their identity and possibly lead to a rejection of the facts presented. However, this approach has certain limitations. First, this process is usually carried out in an institutionalized manner and performed by experts and sources close to journalism. At the same time, the volume of false information online is constantly growing. The vast majority of information (true and false) on social media is not and will never be checked. This therefore reflects a scaling issue. In addition, some users expressing a lack of knowledge of the method may show an unfavorable attitude toward fact-checking, which generates distrust.

In 2023, the Duke Reporters' Lab counted 417 active fact-checking projects in more than 100 countries around the world and in 69 languages. The Poynter Institute, a nonprofit journalism school and research organization, has established a global network of fact-checkers, journalists and ambassadors (public figures, experts, influencers, etc.) with the aim of pushing back on the mass dissemination of false information. Recently, more and more corrective initiatives involving virtual communities are emerging.

Since 2021, X has set up a process first called "Birdwatch", which we know today as "Community notes". This allows users to report posts that they consider to be misleading and, most importantly, to add comments for corrective and informative purposes describing the real context surrounding the information in question. These notes are then visible to other community members who can vote on the perceived usefulness, truthfulness and credibility of the correction. Similarly, WhatsApp has partnered with 50 organizations exclusively dedicated to fact-checking. Network users who receive information that they consider potentially misleading can forward it by direct message to these organizations. These then respond with an appropriate correction, which users can then in turn pass on.

The proliferation of this type of collaborative approach and the increase in exchanges between stakeholders involved in the fight against false information seem encouraging for the future of fact-checking. This would make it possible to overcome the main limitations of the process as currently implemented. The more people committed to verifying and promoting the facts, the more information will be analyzed. This would also streamline communication with ML models and help to accelerate improvements in their ability to detect misleading information. Also, the variety of actors involved should make it possible to reach more users and induce a feeling of confidence among those who have previously felt a certain reluctance.

6.4.2. *Preventing the spread of the disinformation virus: inoculation and media literacy*

To effectively combat the spread of false information on social media, other methods complementary to fact-checking are deployed. These intervene earlier in the process of information dissemination, since they act before exposure to potential fake news and therefore have a preventive aim (prebunking). Among these approaches, we find media literacy interventions and inoculation.

Media literacy can be defined as the ability to analyze, evaluate, communicate and access various media messages (Maksl et al. 2015). It allows individuals to understand how the media industry works and the effects that broadcast messages may have on them. In the context of social networks, this leads them to adopt critical and proactive consumption of media and information, since they are no longer only consumers, but also become creators of content. Generally, users with a high level of media literacy can understand information in a complex way and recognize its source, identify facts and analyze its credibility, which appears particularly useful in the context of fake news. Media literacy also reduces support for (and adherence to) conspiracy theories. It encourages the development of critical thinking on social networks. Today, many organizations on the Internet offer initiatives for media literacy education. This is the case, for example, with UNESCO which offers tools and methods on its website. The institution offers training programs accessible to Internet users. The Poynter Institute has an entire division dedicated to media literacy: MediaWise. The aim of this initiative is to offer several means, aimed at various target audiences (adolescents, seniors, etc.) and by appealing to different actors (journalists, experts, influencers, social networks, etc.), to raise awareness among the population and combat online misinformation. Their programs include a variety of content offering: online courses, educational and animated videos, a campaign aimed at seniors, virtual events (conferences, master classes, etc.), partnerships, media collaborations (partnerships with major economic players, such as Meta, TikTok and Google), a network of ambassadors and face-to-face training modules.

Similar to media literacy interventions, another preventative method has also been developed to raise individual awareness of the dangers of misinformation: inoculation. Regarding this theory, social science follows the medical analogy that exposure to a weakened dose of a pathogen triggers the production of antibodies that provide protection against future infections. For inoculation messages to be effective, they must involve two elements: a generic warning and the example of a fallacious argumentation technique. On the one hand, it is necessary to warn the receiver of the message of the existence of a threat related to an attempt at deceptive persuasion. On the other hand, it is necessary to identify the subterfuge used to mislead the individual. This type of approach is already used in the field of cybersecurity where individuals are regularly warned of attempts at hacking or spam. Currently, many such interventions are designed based on a game model. This method allows the use of active inoculation techniques, which are more effective than passive approaches. Thus, participants themselves will create their own "antibodies" to attempts at false arguments. For example, the *Bad News Game* was developed to teach players, in the form of an interactive tutorial, how to identify fake news online (Roozenbeek and van der Linden 2019). Following the success of the game, researchers from the universities of Cambridge and Bristol teamed up to create the Inoculation Science website, where Internet users can find useful resources. Several other games have also emerged (see Figure 6.2), such as *Go Viral*, *Harmony Square* or even *Cranky Uncle*. These are also used by actors in education. Recently, Google expanded its inoculation campaign to Europe, offering a set of videos dissecting different techniques used by creators of fake news.

To fit the social media format, alternative forms of inoculation or incentives to media literacy could be offered in the form of posts. This type of intervention should be disseminated more easily and on a wider scale due to ease of production and implementation. Unlike games or registering for training courses, posts on the news feed do not require any pre-existing motivation. Furthermore, to counter the potential decreasing effects of inoculation over time, some researchers advocate "booster doses" using the medical analogy. These reminders can be made in the form of a repetition of the initial message (in its entirety or in a summary) or of a new post to renew the impression of threat around certain subjects.

Figure 6.2. *Screenshots, from left to right: Bad News Game, Go Viral, Harmony Square and Cranky Uncle*

In view of these elements, it seems necessary to combine all of the above-mentioned approaches to effectively combat fake news in the smart city. While the use of modern technologies and innovative automated processes such as AI seems promising, these approaches cannot currently operate without manual human intervention. It also seems crucial to intervene at different stages of the diffusion process. This would involve protecting social media users before potential exposure by offering them the tools needed to counter an attempt at fallacious argumentation. Then, to strengthen their resistance, it is essential to verify the facts and point out any inconsistencies and inaccuracies in order to re-establish the truth. In conclusion, the actors participating in the life of the smart city must act in unison to promote critical consumption of the media environment, while maintaining a healthy dose of skepticism, and ensure that a better quality of information circulates.

6.5. References

Batty, M., Axhausen, K.W., Giannotti, F., Pozdnoukhov, A., Bazzani, A., Wachowicz, M., Ouzounis, G., Portugali, Y. (2012). Smart cities of the future. *The European Physical Journal Special Topics*, 214(1), 481–518.

Domenico, G.D., Sit, J., Ishizaka, A., Nunan, D. (2021). Fake news, social media and marketing: A systematic review. *Journal of Business Research*, 124, 329–341.

Gottfried, J. and Shearer, E. (2016). News use across social medial platforms 2016. Report, Pew Research Center.

Grimmelmann, J. (2015). The virtues of moderation. *Yale JL & Tech*, 17, 42.

Maksl, A., Ashley, S., Craft, S. (2015). Measuring news media literacy. *Journal of Media Literacy Education*, 6(3), 29–45. doi: 10.23860/jmle-6-3-3.

Rainie, L., Funk, C., Anderson, M., Tyson, A. (2022). AI and human enhancement: Americans' openness is tempered by a range of concerns: Mixed views about social media companies using algorithms to find false information. Report, Pew Research Center: Internet, Science & Tech [Online]. Available at: https://www.pewresearch.org/internet/2022/03/17/mixed-views-about-social-media-companies-using-algorithms-to-find-false-information/ [Accessed 11 July 2023].

Roozenbeek, J. and van der Linden, S. (2019). Fake news game confers psychological resistance against online misinformation. *Palgrave Communications*, 5(1), 65.

Talwar, S., Dhir, A., Kaur, P., Zafar, N., Alrasheedy, M. (2019). Why do people share fake news? Associations between the dark side of social media use and fake news sharing behavior. *Journal of Retailing and Consumer Services*, 51, 72–82.

Vaccari, C. and Chadwick, A. (2020). Deepfakes and disinformation: Exploring the impact of synthetic political video on deception, uncertainty, and trust in news. *Social Media + Society*, 6(1), 205630512090340.

Visentin, M., Pizzi, G., Pichierri, M. (2019). Fake news, real problems for brands: The impact of content truthfulness and source credibility on consumers' behavioral intentions toward the advertised brands. *Journal of Interactive Marketing*, 45, 99–112.

Vosoughi, S., Roy, D., Aral, S. (2018). The spread of true and false news online. *Science*, 359(6380), 1146–1151.

We Are Social and Hootsuite (2023). Digital 2023 [Online]. Available at: https://wearesocial.com/fr/blog/2023/01/digital-2023/ [Accessed 10 July 2023].

Westerlund, M. (2019). The emergence of deepfake technology: A review. *Technology Innovation Management Review*, 9(11), 39–52.

Yang, K., Varol, O., Davis, C.A., Ferrara, E., Flammini, A., Menczer, F. (2019). Arming the public with artificial intelligence to counter social bots. *Human Behavior and Emerging Technologies*, 1(1), 48–61.

PART 2

The Smart City: What Lifestyle for Citizens?

Introduction to Part 2

The smart city is generally envisaged from an outsider's perspective, as an object of study and action for multiple stakeholders, elected officials, city technicians, large digital or utility companies, in particular (energy, water, waste, etc.). However, to understand the city and the way in which it is changing, we must take everyday lives into consideration by trying to adopt an insider approach, looking at the city through the eyes, usages and practices of inhabitants, in terms of consumption, public services, culture, etc. To move away from an overarching vision, introspective and reflective work is needed to account for the ways in which people inhabit the city (Chapter 7), feed themselves (Chapter 8), educate themselves (Chapter 9), consume (Chapter 10), move around (Chapter 11), travel (Chapter 12), take care of themselves (Chapter 13) and learn and train (Chapter 14).

As Nicolas Merveille points out in the conclusion of this book, the purpose of a city project is to preserve quality of life, to ensure the well-being of a society, and not to build a digitized city. In smart city projects, technology must be seen as a means and not an end in itself. It can nevertheless change people's lives as it becomes part of daily practices and uses (mobility, purchases, waste management, health, etc.). We are all witnesses to, actors in, or victims of this accelerated penetration of digital technology into our daily lives for making medical appointments, organizing our meals, managing our home, entertaining ourselves, etc. This acceleration comes from available technology that is constantly being updated and progressing in line with scientific and technological advances. It also arises from an increased demand for technologies that simplify daily life by saving time or money, providing easier access to a multitude of resources and leisure activities, or working without having to commute (i.e. the explosion of remote working since the Covid-19 crisis, for example).

Introduction written by Gilles N'GOALA.

In this way, technology can contribute to the development of a "minimum effort" society, where everyday tasks (booking tickets to shows or museums, closing the shutters of your house, sorting and dropping off waste, preparing meals, taking the next tram into town, meeting with colleagues, learning at university, etc.) would result in the lowest possible expenditure of energy, time and money. Smartphones have thus become one of the most important and addictive objects in everyday life. By offering everything on the same device, the smartphone has become an essential support for services and a significant element in establishing a service-based city. By offering to maximize pleasures and minimize penalties, digital technology opens up an advantageous path that few obstacles can obstruct, apart from attempts at regulation by states and local authorities (AI, personal data, limitation of uses, etc.).

The somewhat provocative term "smart life" demonstrates how we sometimes lack reflexivity about our own lives. Intelligence refers, on the one hand, to our abilities to know, understand and discern with clarity the world around us, and, on the other hand, to our capacity to adapt to new situations and to independently find solutions to deal with them. Referring to a civilization of goldfish, Patino (2019)[1] deplored how the attention economy had led us to a form of voluntary servitude, new forms of addiction and a society giving primacy to passion (emotions and affect) over reason (information, critical thinking and understanding). As the following chapters show, however, the picture of the digital transformation of our lives is not so bleak. First, our lives are not limited to technological uses. Most of our life experiences are based on human relationships (dinner with family, chatting with friends, etc.), not digital mediations or tools. Second, digital technology is the subject of many debates, within families, companies, territories and countries, attesting to the vitality of the critical thinking and sometimes the resistance of part of the population. Chapter 5 on social acceptability and Chapter 6 on information literacy highlight the proliferation of these debates, but also the importance of a neutral and credible scientific voice on which the population can rely. Finally, intelligence is also the ability to adapt to a very rapidly changing environment. However, the many examples used in the following chapters highlight people's potential inabilities to manage their connected environment, autonomously improvise new solutions for consumption and mobility and integrate societal issues into their daily practices (climate emergency, overconsumption, fight against inequalities, etc.).

Consequently, the rise of digital technology poses questions for humans and their ambivalence toward the objects and technological artifacts around them. Humans are dominated by their technological environment as much as they dominate it.

1 Patino, B. (2019). *La civilisation du poisson rouge : petit traité sur le marché de l'attention*. Éditions Grasset, Paris.

Digital transformation leads to many paradoxes (N'Goala 2016)[2] and cannot fully meet all of the challenges that arise, notably those of responding to the climate emergency and preserving individual and collective well-being in cities facing a crisis of growth.

The city is also a model that is constantly questioned, subject to numerous criticisms, and fuels both our topophobia (insecurity, pollution, noise, overconsumption, cost of real estate, traffic jams, etc.) and our topophilia (distractions, dynamism, youth, productive activities, etc.). This may explain, in France, the development of peri-urban areas that promote access to the city without having its major inconveniences (INSEE 2021)[3]. This work deals with the smart city – and not with smart territories in general – because it is cities that benefit most from a strategy, infrastructure and dedicated resources. It is also cities that concentrate most of the production and consumption of resources and the negative externalities of growth (demographic, economic and technological). However, digital technology affects all territories and impacts daily life. The following chapters will analyze and question technologies in terms of their functions, that is, their role, usefulness and contribution to the life of society as a whole. They transform our lives, positively or negatively, and integrate all aspects of our society (culture, tourism, consumption, mobility, health, food, etc.).

2 N'Goala, G. (2016). Le marketing dans un monde connecté, un monde de paradoxes. *Décisions marketing*, 84(4), 5–18.

3 Alessandro, C., Levy, D., Regnier, T. (2021). Une nouvelle définition du rural pour mieux rendre compte des réalités des territoires et de leurs transformations. Report, INSEE, Montrouge.

7

Inhabiting

7.1. Introduction: everyday life in the smart city

Inhabiting can be defined as living in (a place), as people or animals do. This short definition relates to the different dimensions developed in the literature: the action of occupying, in other words, the use, the habit or the quasi-daily occurrence of this use, and the place which is the object of the use. Throughout this chapter, we examine and apply this idea of living within cities described as smart. Smart living is, in fact, a central dimension of the fields of application of the smart city according to the model of Giffinger et al. (2010). According to these authors, the use of digital technologies in this field is not only intended to provide everyday quality housing, but also to consider social cohesion, security, accessibility to public services (health, education and culture) or tourist attractiveness. The concept of habitation encompasses a multiplicity of facets. This complexity sometimes makes urban planning projects and public policies difficult to implement and requires long-term thinking. This must come from a vision that is itself shared and formulated according to what the stakeholders envisage for the future. In a context where health, geopolitical or economic crises are almost constant and where numerous challenges caused by climate change and the destruction of biodiversity are inevitable, it is very difficult to project ourselves several decades from now. This is not only a difficult job for communities, but also difficult for their residents. Several social science studies deal with individuals' perception of the future. For the literature, two dimensions are to be considered: the probability that the scenario will occur and its desirability. It also emerges that individuals' predictions and expectations generally result from a transposition of their perception of the present context into a future constructed from their imagination. The more distant the future, the more necessary the appeal to imagination. But how can we imagine this future? If imagination is an individual faculty, the imaginary is a construction arising from

Chapter written by Sarah MUSSOL and Pauline FOLCHER.

socio-historical antecedents, like cultural symbols, shared by a community (Martin 2004). This social imaginary (Castoriadis 1975) of what living will be like in the future is interesting to deploy here.

When we consider this future and the social imagination related to it, we very quickly notice the importance of cultural and artistic productions in constructing these shared symbols and representations. How can we consider the digital future without necessarily evoking *1984* by George Orwell or using images from films like *Blade Runner* or *The Fifth Element*? These dystopian creations, fundamental to popular culture, presented the image of a smart city that is not only highly technological, where humans must coexist with robots, but also conveyed an imaginary rich in new forms of habitat. Starting from this idea that today's social imagination of the digital future influences individuals' representations and attitudes (whether citizens, urban planners, politicians or architects), we will first explore representations of habitat and habitation in the smart city in culture and the media. Second, we will discuss current discourses by elected officials on housing in the smart city. Is it more ecological, collective, technological? To what extent does it meet the expectations of residents who certainly want ecology, but also comfort and effortlessness? Finally, we consider the future of housing in its articulation and integration into the neighborhood and into the city, within arrangements providing services and value for residents.

7.2 Inhabiting tomorrow: representations of housing in the smart city

As we mentioned in the introduction, the social imaginary is particularly interesting to mobilize to study and understand modern society (Castoriadis 1975; Martin 2004), and especially representations of a more or less distant future. As we know, in the social sciences, studying representations often allows us to identify and understand beliefs, attitudes and sometimes even behaviors and decision-making. We distinguish two forms of representations: social representations and conveyed representations.

Social psychology, but also marketing, through its study of consumer behavior, takes a particular interest in social representations, defining them as "a form of socially developed and shared knowledge, having a practical aim, and contributing to the construction of a reality common to a social group" (Jodelet 1989, p. 53). These social representations, sometimes forming a social imaginary, are then influenced by images, speeches or, more broadly, stories that eventually form what we call here conveyed representations. When it comes to questioning social representations of the future, the representations conveyed by fictional stories are at the heart of the formation of beliefs and attitudes.

In this section, we propose to return to representations of the city and of the life of tomorrow through the stories of cultural productions, such as science fiction films.

7.2.1. *Stories of inhabiting tomorrow*

Transmitters of these stories of the digital future are diverse and take various forms. The *mass media* are first to come to mind. From interviews with experts or based on (sometimes, distorting) scientific articles, the media seek to anticipate and glimpse this future based on today's reality. "What will the cities of the future look like?" was the headline in *Le Monde* in 2020, or "2041: welcome to the city of the future" for *Les Échos* in 2021. However, we must not neglect the importance of narratives developed around the imaginary, notably with the help of fiction. There are many advertisements or cultural works which seem to predict the future. As some are broadcast on a large scale, these productions may have an influence on our collective unconscious. Indeed, several marketing studies (Adaval et al. 2007; Van Laer et al. 2014; Touzani and Hirschman 2019) demonstrate the role and influence of fictional narratives on affective and cognitive dimensions, as well as on the formation of beliefs (Pereira da Camara and Folcher 2023). The narrative transport model, developed by Greimas (1966) and modeled by Green and Brock in 2000 (*Transportation-Imagery Model*), notably helps explain the power of narrative over the construction of a shared imaginary.

Many works of science fiction, whether through novels (*1984* by George Orwell or *Brave New World* by Aldous Huxley), films (*Blade Runner*, *Minority Report*, or *The Fifth Element*) or series (*Black Mirror*, *Silo*, or *Foundation*), sustain the images of this digital future. Advertisements seeking to promote connected objects or voice assistants also feed this imaginary of digital everyday life and technological solutionism, which we find in the myth of artificial intelligence and technical progress. Finally, if we are not really able to understand or concretely perceive what it is like to live in a smart and digital city, individuals will tend to turn to what they know to form an image of this future. This is the point where popular culture can permeate and influence shared beliefs and attitudes toward the smart city.

Cultural productions, such as science fiction films, will, through narrative transport, mainly convey representations of a rather dystopian digital future (Pereira da Camara 2022; Pereira da Camara and Folcher 2023). Based on this work, we can identify several elements of the representations conveyed through science fiction films. Chronological analysis of cinematographic productions first allows us to highlight the evolution of directors' evocation of the future.

> If in the 1980s the plot takes place in a distant and very futuristic period, productions of the 21st century tend to take place in the near

future in which the technology is implemented in contemporary networks and urban settings.

In addition, the authors highlight the importance of the historical context present in representations of governance and of the forms of exercising power. The models proposed are thus deeply rooted in the present, like the emphasis on the power of large companies in the films of the beginning of the 21st century. It is also interesting to look at how technology, originally represented by humanoid robots, gradually becomes omnipotent, while being invisible.

Filmmakers' depictions of what life will be like in the city of the future are generally not very encouraging. Indeed, the image conveyed over the decades is often that of a city ravaged by human error or completely sanitized and subject to totalitarian governance. Violence or alienation from technology are often at the heart of essentially authoritarian modes of governance, which leave little room for collective action and intelligence. In these film productions, there is little room for the living world other than humans and fauna or flora are almost nonexistent, which makes the atmosphere and decor quite gray and lifeless. Citizens only very rarely appear as happy individuals. They are herded into high-density neighborhoods according to their class, ethnicity or language.

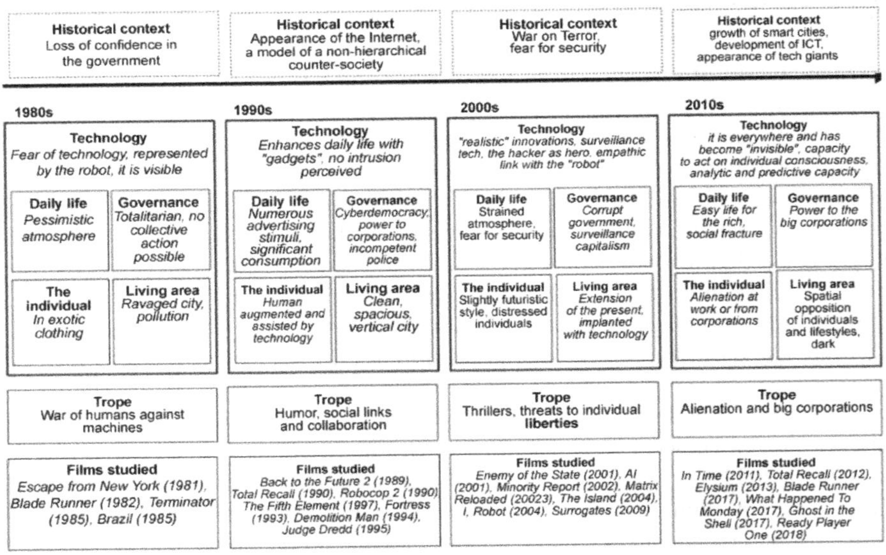

Figure 7.1. *Historical analysis of images of smart cities in fiction (Pereira da Camara and Folcher 2023)*

Technology, for its part, is everywhere, more or less visibly, with the main objective of monitoring citizens for security purposes. Given analytical capabilities, technology is able to anticipate, predict, prescribe and is used by those in power or corporations to control behavior. The pervasiveness of technology in the practices and sometimes the consciousness of the inhabitants seems to be largely accepted, and few of them rebel against the intrusion and permanent capture of individual data. The individual is constantly solicited through technology, with the help of implants and sensors or through advertising stimuli of monumental size seeking to constrain their consumption choices. *Smart people* have little power of decision and action in their consumption. Indeed, behaviors and consumption patterns, even in the most recent films, are only very little modified (except when food is a rare commodity), or, if they are, they become even more frenetic and controlled by firms or technology.

We also note that if climatic peril and the habitability of the planet are featured, often at the beginning of the plot, the consumption behavior of individuals is not represented as more sober and virtuous. Smart objects, and more broadly the technology used, are, most of the time, used with the goal of utility and individual satisfaction (particularly for the richest social groups).

Finally, major social questions, such as the debate between collective interest versus self-interest and collective security versus individual freedom, also underlie the narrative, but the people are shown as submissive and having little confidence in collective efficacy. The conception of smart people, as it is developed in the literature on the smart city (Gupta et al. 2017), is thus very far from the image that science fiction films have made of it for decades (Pereira da Camara 2022; Pereira da Camara and Folcher 2023).

7.2.2. Representations of the habitat of the future

Representations of housing in cultural productions, such as novels, comics or science fiction films and series, often highlight technological advances and futuristic concepts, aiming for the sake of the plot to create imaginary worlds, where housing is often more advanced, innovative and sometimes even quite futuristic.

The architecture of buildings and housing is often futuristic, with bold shapes, clean lines and innovative materials. Structures may be gigantic, rising into the sky or even floating in space. Verticality conveys both the problem (this time very real) of overpopulation in cities, where inhabitants are crammed on top of each other, and a certain social segregation, where the rich and powerful live ever higher and the poorest ever lower. It is no longer a question of social differentiation by neighborhood, but by the vertical position of the housing occupied.

As in *The Fifth Element* or even *Back to the Future 2*, homes, in their futuristic representation, are often equipped with particularly advanced home automation equipment. Characters thus control their homes using voice or gesture commands (no more typing on a keyboard) to adjust brightness, lighting, temperature, surveillance or any other aspect of their environment. Sometimes a voice represents the concierge-assistant interface, answers questions, gives recommendations or acts more or less autonomously; a little like what Siri, Alexa or Google Home aspire to today. These assistants are found alongside the protagonists not only in the home, but also outside.

Like this voice, futuristic homes often seamlessly integrate technology into their design. Touch screens are embedded in walls or windows, can be transformed into display screens and everyday objects are equipped with advanced features. The protagonists never question the intrusiveness of the technology and hardly worry about the processing of their personal data and respect for their private life. Technological solutionism is promoted and connected objects are mainly presented as facilitators of everyday life.

All this means is that these homes are often presented as intelligent and capable of adapting to their occupants' needs. For example, the habitat can automatically adjust lighting and temperature based on each person's preferences, or even anticipate occupants' needs and desires. It is no longer a question of smart objects alone, but rather of an ambient intelligence, where the home captures, compiles and interprets data for autonomous action.

Finally, we may note that housing is generally presented in its purely functional dimension, sometimes taking the form of a capsule. It is only the place where the protagonists sleep or refuel. It rarely functions as a home. This symbolic dimension of the habitat is then ignored, sometimes for plot reasons, and where it is present, it is above all because the characters are confined there.

To conclude, the representations conveyed by fictional productions are often exaggerated for the needs of the plot or emerge from the directors' fantasies. At the same time, they question the architectural and technological possibilities that we would like to implement. Furthermore, how can we imagine that all of these images conveyed of this city and this life of the future do not influence each of our perceptions and mental representations? How can we deny the fact that watching *Minority Report* or reading *1984* can influence beliefs about policies of security and policing assisted by cameras, data and digital technology in general? While the influence of the representations conveyed by cultural productions is obvious, the question arises as to how no longer filmmakers, but decision-makers, politicians and urban planners envisage the city, the neighborhood and housing in the city of tomorrow.

7.3. Political visions of inhabiting tomorrow

Political speeches and public presentations of (current and future) development projects by local authorities are interesting to dissect and analyze, to identify trends, visions, or, as they say in marketing, the value propositions proposed by local authorities to their residents. In France, two major challenges in particular concern communities: ecology and urban densification. Generally speaking, five major objectives guide public housing and habitat policies: sustainability and energy renovation, management of densification, accessibility of housing and social diversity.

Thus, in Montpellier, the PLUI-Climat (Local Plan for Intermunicipal Urbanism-Climate), proposed in 2023, meets two objectives: the reduction of the area to be urbanized and densification along transport routes. The question of density is central as it provokes strong reactions from citizens, but nevertheless allows us to meet ecological objectives. The Citizens' Convention for the Climate therefore formulated the following recommendation: "raise awareness of the importance and interest of a more compact city and build a new culture of collective housing".

We have seen that science fiction scenarios tend to overlap on a number of points that sometimes seem dystopian to us. However, these scenarios, by creating fictional attentions (Beckert 2013), tend to shape our expectations for the future. Elements such as the omnipresence of the car in the city (*The Fifth Element* or *Back to the Future*) are thus quite far removed from society's current concerns, for example, on ecology. Political discourse may propose alternative scenarios.

7.3.1. *Political discourse of the smart city*

The decision to develop a smart city project is, of course, political. Whatever the motivations and operational objectives of these projects, they are also used to convey a vision of the territory focused on innovation and the understanding of this future. It is then interesting to identify the elements of language, and more broadly the discourses and representations of those who "make" smart cities, especially in France.

Through a qualitative and thematic analysis of comments made by elected officials and smart city project managers in press articles (both in print and on the web) published in France in national and regional media in the first half of 2019, we can thereby recover the political expectations and hopes associated with these projects.

In these speeches, the smart city is above all presented as a management tool (19% of the corpus) available to business departments whose goal is the better performance of processes and greater efficiency in decision-making. The examples given are often the same and revolve around four main areas of the smart city, namely, mobility, the environment, security and the economy. As the following verbatim quote states that, "it allows us to go further in energy management for buildings". Emphasis is also placed on the configuration, the technical and digital networking implemented by cities. "We will convert the 34,000 streetlights to LED, and we will equip them with sensors". The user and services are, for their part, presented overall in terms of the proposed applications, digital devices, more than by the value that this will bring to daily life. Finally, the political dimension is confirmed here. Statements that recount pride in innovating, in being the first, represent 18% of the corpus and can be illustrated by the following verbatim quote: "I never imagined, four years ago, that I would be launching a unique smart city project which would be competing with that of Singapore".

7.3.2. *Prospective scenarios for the habitat of tomorrow*

Prospective scenarios agree when, based on scientific reports such as those of the IPCC, they envisage significant upheavals in lifestyles in the near future. Increasingly rapid technological innovations and increasingly urgent climate issues will undoubtedly change the daily lives of all ecosystems, and even more so the urban ecosystem. Demographers and geographers estimate that in the coming years, cities should become significantly denser. With this phenomenon, the problems of urbanism or public management linked to housing needs and the management of flows (human and goods) will also intensify. Service provision, whether public or commercial, will also have to change not only in terms of inclusive configuration (accessibility, diversity, quantity, dynamics and acceptability), but also from the perspective of a positive impact of general and environmental interest, notably by promoting circularity and collaboration. The challenges of habitat, but above all of habitation, are paradoxically both central in the short term for residents and part of the long term, regarding the realization and development of projects. Always at the heart of political projects, the neighborhoods of tomorrow are taking shape today.

In the design of these neighborhoods of tomorrow, often developed as eco-districts, the stated ambition of municipalities is the development of livable cities for residents in the relatively near future. Thus, although ecological objectives are announced, they are often relegated to the background, in favor of laudable social ambitions. Thus, in Nantes, the Bottière-Chénaie eco-district project aims to limit urban sprawl. "This new district was designed to primarily meet the expectations of low-income households and young workers with an affordable housing supply, so that they can continue to live in Nantes". In this type of eco-district, accessibility is the first consideration.

The question of urban sprawl is often combined with that of mobility in these neighborhood projects. In Rennes, "the La Courrouze neighborhood is inventing a new urban territory. Launched in the 2000s, its development will continue until 2028". By bringing together diverse housing, tertiary activities, shops, services and facilities, the neighborhood helps to create a short-distance city. The new neighborhood facilitates exchanges, like a "city link" which opens up and integrates the southwest sector into the agglomeration. Enhancement of natural heritage is at the heart of the project. The development takes special consideration of the ecological continuities across the La Courrouze site that are included in the large urban park, which constitutes the neighborhood. A park punctuated by a diversity of atmospheres.

In several cities, participatory housing projects are being realized, such as the Cartoucherie neighborhood in Toulouse (social/intergenerational mix). For metropolises that are large in terms of surface area, the decentralization of the city into self-sufficient neighborhoods logically leads to a reduction in the need for motorized travel by residents and, in fact, congestion on the roads. The positive impact on air pollution and biodiversity allows communities to meet the climate goals they have set for themselves. For these same cities, rehabilitation of infrastructures and the development of third places make it possible to design attractive habitats of the future.

In a context that is both post-pandemic and includes diplomatic and economic tensions and significant societal transitions, a neighborhood is no longer just the place where housing is located, but it is the space where we live. We live there, become involved there, work, have fun, play sports or get supplies and meet our loved ones every day. These new uses by residents (ubiquitous for many) lead to a rethinking of the supply and production of local services.

7.3.3. *Political discourse as the basis for political action*

Urban planning researchers agree that "a high degree of decentralization was favorable to the bottom-up initiatives upon which the sustainable city and sustainable mobility are often based" (Emelianoff 2005; Tortosa 2023). Studies and examples show that stakeholder participation is even more important when they are involved at the local, or even ultra-local, level, in their neighborhood, their street or even their building. Experimentation with innovative projects is also essential for their development and the small scale offered by the neighborhood is particularly interesting. With this refocusing on immediate proximity, the city is now more than ever becoming a constellation of neighborhoods, each with both its own identity and singularity, offering an inclusive and harmonious configuration of infrastructures and services across the urban area as a whole, seeking to develop citizen participation and involve residents in decisions and projects taking place near their homes, to reduce road congestion and travel, to promote population mixing and social diversity, etc. The objectives are diverse and vary according to the context and situation in which the territories find themselves.

Norway is an example of this. Indeed, this country has been thinking about its policy of reducing car travel since the early 1990s (Tortosa 2023), where, "under the impetus of the state, local authorities are then invited to adopt a new urban planning policy which reduces the need for travel, contains urban sprawl, and densifies housing" (Tortosa 2023). The short-distance city is thus the guiding principle for urban planning projects. At the same time, provision of public transport services and the promotion of soft mobility are improving to offer an alternative to the car and to ease congestion in Oslo, in particular. Today, the range of measures is completed by urban tolls in large cities. In his article published in 2023, Tortosa looks back at the results of this policy and explains to us that:

> [If] at the end of the 1990s, the modal share of the car was of the same order in large cities as across the entire country, there was a subsequent decoupling, with a much more marked decline in the former, particularly in Oslo and Trondheim where over the 1998–2018 period the decline was 49% and 22% respectively. National measures to restrict cars have therefore borne fruit.

Still in the perspective of developing more sustainable territories, the adoption of a system based on circular processes seems to be increasingly shared by communities in many countries. A circular vision implies thinking about the production and consumption of energy flows in their entirety across the territory, seeking to optimize and adapt each stage of the process in order to meet individual needs, while seeking as much as possible to reduce its impact on the environment. In these urban projects, buildings are designed not only in line with the nature of the soil and the sunlight, but also their role in the life of the neighborhood alongside other infrastructures. In its "City Project: Towards an Urban Planning and Mobility Plan", the urban planning team imagines "buildings that no longer require energy to be heated in winter or air-conditioned in summer. The orientation of buildings, lighting and natural ventilation, and the choice of materials are solutions suggested for adaptation of buildings", or again: "rain gardens and permeable soils allow for on-site rainwater management".

However, given the complexity of tomorrow's challenges and the speed of the changes which cities will have to face, political discourse may still lack clarity. Beyond the choices in urbanization highlighted above, political discourse also aims to involve citizens in major changes. Thus, as Emmanuelle Wargon, Minister Delegate in charge of housing in France from 2020 to 2022, points out, "the political vision of city planning and urban innovation often lacks clarity for the French, even though many steps have been taken to encourage territories in a virtuous logic, promote exemplary projects, or disseminate knowledge".

7.4. Residential practices in the smart cities of tomorrow

For households, choosing a home is sometimes a complex process that requires implementation of a residential strategy. The latter is defined as an acceptable compromise between the constraints (of the housing market, individual economic and financial constraints) and the objectives set by households (share allocated to housing in the household budget, proximity to the workplace, to family, neighborhood tranquility, proximity to services, etc.). So, while the French dream remains the individual house, this does not fully respond to most of the challenges linked to climate change.

To study the trade-offs made by individuals in their choice of housing, researchers in urban planning and urban sociology at the end of the 1980s developed a framework of analysis based on three dimensions: occupation status (owner/tenant), type of housing (individual house, apartment, gated community, collective or intergenerational housing) and location (city center, fringes and residential area).

If this framework is still relevant today, the question then arises as to how these residential strategies and practices evolve and how smart city projects conceive of housing as such and how they integrate it into the city. With the increase in population density in the heart of cities, the trend is toward verticalization and development of apartments (now more numerous than individual villas in most cities). For information and comparison, in France in 2021, single-family homes represented approximately 57% of principal residences, while apartments represented approximately 43%. In Paris, the largest urban area, the figures were quite different, since apartments represent more than 96% of principal residences (INSEE figures for 2018). In Quebec, across the province, the figures are almost equivalent to those in France with 60% of single-family houses. In Montreal, the figures are more nuanced than in the French capital, since there are still 30% of individual houses.

Note also that, at the same time, the *smart home* market (associated with the "smart city") is strongly developing and represents more than three million dollars of revenue in 2023. Thus, more than seven million homes feature this technology in France. The smart home can be thought of as a home that integrates digital sensing and communication devices (Gram-Hansen and Darby 2018). The smart home may thus include one or more voice assistants, energy and consumption management equipment (electricity, gas and water), monitoring equipment and/or connected household appliances.

Although we say *home*, the vast majority of these programs are apartments. Indeed, in smart city and/or neighborhood projects around the world, it primarily means residential buildings equipped with sensors, IoT and home automation systems. The objective is then to use different technologies to know and understand the habits of the inhabitants, to make recommendations to them based on their objectives and thus move toward more sustainable consumption, particularly energy consumption. But these smart buildings are also equipped with external sensors allowing adaptation and optimization of the building's consumption, for example, by capturing the light and heat of the sun. Some are even equipped with solar panels, to be energy independent, or use plants and green walls to cool spaces and combat heat waves in city centers.

Although, following the health crisis of 2020, many French people aspire to a peaceful life far from cities, the future of housing seems to be primarily in the heart of dense cities, in which ecological issues can be considered. The rise of remote working has also highlighted the need for functional layouts in housing, and modularity of spaces is increasingly becoming the new "norm" in our way of living.

7.5. References

Adaval, R., Isbell, L.M., Wyer, R.S. Jr. (2007). The impact of pictures on narrative-and list-based impression formation: A process interference model. *Journal of Experimental Social Psychology*, 43(3), 352–364.

Aulagnier-Castoriadis, P. (1975). *La violence de l'interprétation. Du pictogramme à l'énoncé.* PUF, Paris.

Beckert, J. (2013). Imagined futures: Fictional expectations in the economy. *Theory and Society*, 42, 219–240.

Emelianoff, C. (2005). Les agendas 21 locaux : quels apports sous quelles latitudes ? *Développement durable et territoires*, Dossier 4. doi: 10.4000/developpementdurable.532.

Giffinger, R. and Gudrun, H. (2010). Smart cities ranking: An effective instrument for the positioning of the cities? *ACE: Architecture, City and Environment*, 4(12), 7–26.

Gram-Hanssen, K. and Darby, S.J. (2018). "Home is where the smart is"? Evaluating smart home research and approaches against the concept of home. *Energy Research & Social Science*, 37, 94–101.

Green, M.C. and Brock, T.C. (2000). The role of transportation in the persuasiveness of public narratives. *Journal of Personality and Social Psychology*, 79(5), 701.

Greimas, A.J. (1966). Éléments pour une théorie de l'interprétation du récit mythique. *Communications*, 8(1), 28–59.

Gupta, S., Mustafa, S.Z., Kumar, H. (2017). Smart people for smart cities: A behavioral framework for personality and roles. In *Advances in Smart Cities*, Kumar Kar, A., Gupta, M.P., Vigneswara Ilavarasan, P. and Dwivedi, Y.K. (eds). Chapman and Hall, New York.

Jodelet, D. and Moscovici, S. (1989). *Folies et représentations sociales*. PUF, Paris.

Martin, B.A. (2004). Using the imagination: Consumer evoking and thematizing of the fantastic imaginary. *Journal of Consumer Research*, 31(1), 136–149.

Pereira da Camara, S. (2022). Les smart people et le transport narratif dans les films de science-fiction des années 80 à aujourd'hui. Thesis, Université de Montpellier, Montpellier.

Pereira da Camara, S. and Folcher, P. (2023). Chez vous la ville du futur c'est plutôt... Minority Report ou Blade Runner ? Récits cinématographiques et représentations de l'expérience de vie dans un environnement connecté : exploration de 4 décennies de films SF. In *Congrès de l'Association Française de Marketing*, Vannes.

Tortosa, G. (2023). En Norvège, des grands ensembles attentifs aux enfants. *Métropolitiques*, 9 February [Online]. Available at: https://metropolitiques.eu/En-Norvege-des-grands-ensembles-attentifs-aux-enfants.html.

Touzani, M. and Hirschman, E.C. (2019). Islam et idéologie au cinéma : prototypes, stéréotypes et économie politique. *Recherche et applications en marketing*, 34(2), 6–25.

Van Laer, T., De Ruyter, K., Visconti, L.M., Wetzels, M. (2014). The extended transportation-imagery model: A meta-analysis of the antecedents and consequences of consumers' narrative transportation. *Journal of Consumer Research*, 40(5), 797–817.

8

Eating

8.1. Introduction: creating a smart food environment, the case of the city of Montpellier

After having been deprived of food sovereignty for a number of centuries (Daviron et al. 2017), cities are once again investing in territorial policies, in which food supply is becoming a central concern. Many of them are now actively engaged in implementing actions promoting and facilitating access to a healthy and sustainable supply of food. The *Milan Urban Food Policy Pact*, launched in 2015 by the City of Milan, to which 200 cities around the world are now signatories, constitutes one example among the myriad of initiatives that illustrate this reappraisal of the question of food within urban spaces.

Indeed, cities today are confronted with major challenges, whether demographic, economic, social, health or environmental. How can we make food accessible to 6.5 billion people concentrated in only 2% of the planet's land area by 2050 (Testard-Veillard 2022)? At the same time, situations of food insecurity have worsened since the 2020 health crisis. A DREES study shows that food distribution increased significantly between 2020 and 2021, particularly in large French cities. This result, which is not specific to France, has been observed at a global level (Parek et al. 2021; Kakai et al. 2022). From a health perspective, urban areas have been identified for many years as places that may promote poor eating habits (Calise et al. 2019). Territories of obesity (Dupuy et al. 2011) or obesogenic environments (Lake and Townshend 2006) have been observed. At the environmental level, urban areas are the zones that produce the most food waste, with organic waste constituting more than half of total urban waste (Kaza et al. 2018; Fattibene et al. 2020). Finally, there is the carbon footprint of food, within which cities represent the major source of emissions at the global level (Goldstein et al. 2017).

Chapter written by Karine GARCIA, Andréa GOURMELEN and Angélique RODHAIN.

In view of the many challenges posed by urban food supply, this chapter aims to evaluate the potential characteristics of a smart city's food environment that would meet the challenge of accessibility to healthy food for all.

Indeed, according to Albino et al. (2015), the development of a smart city is based on a network approach as well as an urban policy that aims to promote inclusion of its inhabitants and understand the natural environment as a central component of the territory's strategy. In this context, how can we make a city's food environment smart? How can so-called "smart" cities contribute to improving their residents' access to healthy food?

To do this, after defining the outlines and issues of food accessibility, as well as the concept of the food environment, we will focus our attention on the food policy strategy of the city of Montpellier in the context of the various projects implemented since 2015–2016.

8.2. The smart city on food issues

8.2.1. *Reality*

The notion of "smart city" is often controversial, as it is difficult to precisely define the outlines of this concept (Eveno 2018; Maye 2019). The literature often presents it as a city that aims for service efficiency through decision-making based on data collected in massive amounts from connected devices. The outcome is therefore a service-oriented and techno-centric city (Bernardin and Jeannot 2019). This vision may be explained by the very origin of the concept, dating from the end of the 2000s, a time marked by strong acceleration of digitalization thanks to Web 2.0, the Internet of Things, artificial intelligence and the generalization of smartphones (Kim et al. 2021).

In this context, it is not surprising to note that issues related to food in smart cities are rarely addressed in research or professional publications, except for topics dealing with urban agriculture, waste management (Dey et al. 2022) or transport and routing in cities.

At the empirical level, concerning agriculture, this mainly involves soilless cultivation practices carried out in controlled environments, such as hydroponic vertical farms (Palma Lampreia Dos Santos 2016). In practice, these innovations are still in their infancy and therefore few cities are involved. As for waste management, solutions such as smart bins or intelligent transport systems also remain relatively inaccessible and uncommon.

Consequently, while practices such as those mentioned above are still far from being generalized at the empirical level, the food issue in the context of a smart city poses problems at the conceptual level. Unlike issues of mobility, which occur in public space and for which cities have the competence, food consumption mainly takes place in private environments, whether at home, in food retail outlets, or anywhere else that people eat outside the home. It seems difficult to envisage an implementation of connected devices in public spaces that would allow us to understand, quantify and evaluate food consumption, which would help to guide individuals toward healthier eating habits.

Regarding the smart food city, Maye (2019) points out that a technological solution cannot be the only response. Consequently, this complexity leads to rethinking how the food issue is approached in the context of smart cities, dealing with the particularities of this field and developing innovative approaches that do not necessarily rely on highly connected environments to promote accessibility to healthy food for people living in smart cities.

8.2.2. *The importance of an interest in accessibility to healthy food in smart cities*

8.2.2.1. *Food insecurity: heterogeneous profiles*

Food insecurity has increased considerably, notably in large urban areas. According to INSEE, in 2022, no fewer than four million people had to resort to food aid. This issue affects individuals who find themselves simultaneously in economic poverty and facing various sociocultural and political challenges, which impede their access to healthy and affordable food. In concrete terms, this translates into the need to request food assistance that comes with administrative constraints, limited room for action, the obligation to consume unsustainable food products and constrained distribution of these products (Paturel 2019).

At the national and metropolitan level, 52% of food aid recipients are between 25 and 49 years of age, with an average age of 44 years. The groups involved vary depending on the type of aid. For example, food parcels and solidarity grocery stores are mainly accessed by women. The share of people in single-parent families or living alone is almost similar. In contrast, meal distributions mainly attract men, living alone in most cases. One specific category is that of students, who have joined food aid schemes in massive numbers since the health crisis linked to Covid-19 in 2020, especially in urban areas. Food insecurity among students is nothing new, but it has become more visible since then. Indeed, a study by the Observatory of Student Life (OVE) in 2021 revealed that 38% of students had to restrict their food spending, 18% sometimes suffered from hunger and 16% said they had skipped meals for financial reasons.

The diversity of the groups benefiting from food aid (from families in difficulty to students, including people living alone) highlights the scale of the problem and the need for cities, especially so-called "smart" cities, to think of suitable solutions for each category of the population.

8.2.2.2. *The residential neighborhood: source of inequality in access to healthy food*

Since the 1990s, research in the United Kingdom and North America has focused on the influence of place of residence on eating habits. These studies revealed the existence of inequalities in access to healthy food within a single city, thus giving rise to the concept of "food deserts", as defined by the UK Low Income Project Team in 1990. This term is now used to describe areas where access to quality food is hampered by both physical and financial barriers, mainly affecting populations living in socially and economically disadvantaged neighborhoods.

More recently, studies have shown that residents of economically and socially disadvantaged neighborhoods are overexposed to highly processed foods, resulting in an increasing prevalence of obesity-related risks (Cooksey-Stowers et al. 2017).

Unlike food deserts, residents of these areas do have physical and financial access to local food supplies. However, the problem lies in the quality and the unhealthy nature of this supply. In fact, there is an overabundance of high-calorie foods, rich in saturated fats, sugar, salt and other harmful ingredients, far greater than that of healthy foods.

These findings underline the crucial importance of taking measures to improve the quality of food supply in disadvantaged neighborhoods, while recognizing that physical and financial access are only part of the equation. It is equally imperative to promote initiatives aimed at encouraging healthier food choices. Improving access to healthy food in neighborhoods identified as food deserts or food swamps is a complex but essential challenge to ensuring a better quality of life and health equity for all urban residents.

8.2.3. *Accessibility of healthy eating*

Food accessibility can be defined as the ease of accessing healthy food (Calise et al. 2019). According to the WHO, a healthy diet consists of correct proportions of fruits, vegetables, legumes, dried fruits, and whole grains (e.g. unprocessed corn, millet, oats, wheat, and brown rice), in which consumption of sugar, saturated fats, and salts is limited.

In a given city, accessibility may be understood objectively using the three criteria mentioned by Gordon et al. (2011):

– distance, between a resident and the nearest supermarket. Thus, a limit of 15 minutes on foot, equivalent to approximately 1 km, is generally accepted;

– diversity, corresponding to possible alternatives in the number of accessible stores within a radius of 1 km;

– variety, representing the number of stores that provide access to a variety of products at affordable prices.

The challenge is thus to facilitate physical and financial access to healthy food, which constitutes one of the pillars of sustainable food in cities.

However, objective accessibility of healthy alternatives is a necessary but not sufficient condition for developing good eating habits at an individual level. In fact, diet is also influenced by a set of individual factors, such as eating habits and preferences, education, psychological and sociocultural factors. Thus, the subjective dimension of food accessibility, that is, people's feelings, sometimes makes more sense than the objective dimension, for example, the 1-km radius studied (Wang et al. 2015; Vitman-Schorr et al. 2019; Yamaguchi et al. 2019). This underlines the importance of knowing residents' needs, and of making them actors by involving them, in order to direct their habits toward a healthier diet.

8.2.4. *A food environment at the collective and individual level*

The food environment can be seen as a factor that may negatively or positively influence the health of individuals, since the effects of scarce supply, whether real or perceived, will influence the trajectories of individuals within the city, as well as their decisions regarding food (Hamilton et al. 2019). Swinburn et al. (2013, p. 2) define it as "the collective physical, economic, political and socio-cultural environment, the opportunities, and the conditions that influence people's food and beverage choices and nutritional status". This definition emphasizes the multidimensionality of this environment which is not limited to food in the strict sense, but depends on other factors, such as the geographical location of the place of residence (city center versus outskirts, urban area versus rural area), transport infrastructure, the level of wealth of the geographical area, etc. Individuals' eating attitudes and behaviors are thus influenced by their places of residence, which may therefore have consequences for their nutritional status.

As a result, the issue of the smart food environment seems complex to resolve for cities, because it depends on the population studied. However, one of the objectives of so-called smart cities is to promote access to healthy food for all residents, which may seem illusory. Also, cities have limited knowledge of certain groups who are nevertheless vulnerable, for example, students, because they have no system to support them, unlike children in public schools who have school canteens.

In this context, the smart city aims to provide solutions for all its inhabitants, but faces obstacles in their diversity and the ignorance of some of them. As a result, there is no single solution to push all citizens toward healthy eating. Thus, a differentiated marketing approach (adapting services to the expectations of each category of the population) seems particularly necessary.

However, it involves better understanding of the needs of the different types of residents who live side by side in the city. In the following section, we will illustrate the strategy and actions conducted in this regard by the city of Montpellier, which is very advanced in terms of urban food policy.

8.3. The case of the city of Montpellier

8.3.1. *Context of the city and its food policy*

Montpellier Méditerranée Métropole (3M), which has 499,761 inhabitants, 299,096 of whom live in Montpellier, is among the most attractive urban areas in France. At the same time, the city is characterized by its capacity to attract a significant student population, which represents around 18% of its current population. Despite this dynamism, the poverty rate in the city of Montpellier, which was 26% in 2020, is among the highest in France.

Until the early 2000s, actions taken around food issues within the territory were mainly focused on the crisis that had affected viticulture since the 1990s. However, under the influence of the National Nutrition Health Plan (PNNS), the National Food Program (PNA) and through the active involvement of a community of researchers specializing in the fields of agriculture and food at the local level, food has become a major issue there (Michel and Soulard 2019). This conjunction of elements has generated a dynamic that is political, institutional and academic, which resulted in the creation of an "agro-ecological and food policy" (P2A) in 2015, as Hasnaoui Amri et al. (2020) point out.

Furthermore, this decision also led the city to commit to the Milan Pact, an initiative promoting urban food policies bringing together 200 cities around the world, which aims to establish a sustainable food system at city level, as well as in the 31 municipalities of the metropolis.

Montpellier Méditerranée Métropole (3M) and the city of Montpellier in a few figures

– 31 municipalities.

– 499,761 residents, including 299,096 in Montpellier.

– 50% of the population is under 34 years old.

– One-third agricultural space: 14,200 ha.

– 14 hectares of agricultural land in the metropolitan area.

– 640 farms, including 8.1% organic ones, in the metropolitan area.

– Poverty rate: 18.9% (3M) and 26% (city of Montpellier)[1].

Box 8.1. *Main characteristics of the city of Montpellier*

From this period onwards, the objective has been to understand agriculture and food through the lens of a sustainable food system within the territory. This resulted in a first major action in terms of collective catering in the city's schools, with an integrated approach focused on sustainable food defined by Luc Lignon (director of food policy) as "quality both at the organoleptic and nutritional level, good for health, ethical, with fair and equitable commercial relations, and accessible to the greatest number".

In addition to this founding approach, in just a few years the city has become recognized for its innovations in sustainable food systems.

The first experiment with a "common food fund" in France, based on a social security model, was launched in February 2023. A set of projects, to be integrated into the *Cité de l'alimentation* ("Food City") in 2026, are currently being considered.

This section examines three initiatives undertaken by the Montpellier metropolitan area, as part of a sustainable development approach aligned with the principles of smart cities, as observed in Northern Europe (Eveno 2018). The first is a school catering program implemented in the city's 85 canteens. This foundational act constituted an in situ experiment for the city with an audience whose local stakeholders have competences. The second initiative, a "food city", includes all

1 See: https://www.insee.fr/fr/statistiques/1405599?geo=COM-34172 and https://www.insee.fr/fr/Statistics/1405599?geo=EPCI-243400017.

actions in terms of food accessibility and represents a significant change in scale, because it aims to benefit all the city's residents. Within this vast Food City project, the third initiative presented in this chapter is the "Solidarity Food House" project, whose strategy is still being defined.

> The results of the study presented below are a synthesis of analyses based on various materials that were either collected or created during the research.
>
> These materials include:
>
> – Secondary data, such as documents related to the city's food policy projects (e.g. reports, presentation slides, etc.) and to solidarity food department data (see: http://www.montpellier-alimentation-solidaire.xyz/?PagePrincipale).
>
> – A semi-structured one-on-one interview conducted with the Head of Food Policy for the city of Montpellier (duration: 1 h, May 2023).
>
> – A written transcript of a webinar focused on the Ma Cantine Autrement project (November 2021) (see: https://www.arbe-regionsud.org/19454-webinaire-1h-pour-comprendre-et-agir-ma-cantine-autrementnbspretour-de-la-ville-de-montpelliernbsp.html).
>
> – A written transcript of a brainstorming workshop co-facilitated by academics from the University of Montpellier (authors of this chapter), the Deputy Mayor of Montpellier responsible for Food Policy and Urban Agriculture, and the Head of Food Policy. The sessions were held with student groups from the University of Montpellier.

Box 8.2. *Methodology*

8.3.2. *Ma cantine autrement (MCA): an innovative approach to a familiar target for the city*

In 2015, the city completely rethought its strategy toward the food issue by creating a Food Policy Directorate (DPA). This food governance, "the result of the joint impact of administrative and political agendas [...], brings food into the sphere of local political action [...], and leads to considering the transversality between administrative and technical services that possess competences originally organized around the key missions assigned to a French metropolis" (Michel and Soulard 2020, p. 149).

In fact, within local authorities, collective catering is often linked to competences related to education, services relating to youth and childhood. The creation of this department allowed the city to adopt an overall perspective on food management, encompassing its supply policy and extending to waste management. This involves approaching food from a systemic perspective.

– Three departments: central production unit department, quality/diet department and stewardship/maintenance department.

– 85 school cafeterias.

– 42 leisure centers.

– 22,000 children.

– 15,500 meals served per day.

Box 8.3. *Activities and structure of the Food Policy Directorate*

The initial work, carried out between April and September 2014, consisted of creating a diagnostic report on school catering in the city. A second inventory was conducted between 2015 and 2016, covering all collective catering in the metropolis, with a six-month period dedicated to analyzing issues related to food waste.

This diagnostic, conducted over almost two years across the city and the 31 municipalities of the metropolis, is of crucial importance, because it allows all phases of the year to be included and the problems to be understood in their complexity.

The conclusions of this work showed that the unit cost of a meal in the city's school canteens was higher than the national average in France. This observation also applied to the amount of food wasted, partly due to children's dissatisfaction with the meals served and to overproduction.

For example, up to 45% of fruit may be wasted in the cafeterias studied. In addition, the supply of organic and local food remained limited.

In summary, the diagnostic painted a picture of collective catering that was misaligned with the principles of sustainability, thus underlining the need to completely rethink the city's collective catering strategy. It should be noted nevertheless that the direct management of school canteens by the city represents a significant advantage.

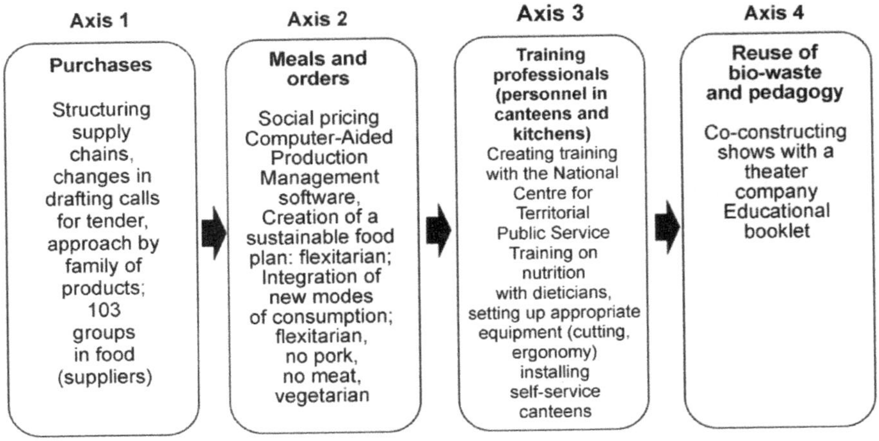

Figure 8.1. *Milestones of actions implemented from 2016 onward under the MCA program*

A common foundation is first defined at the city and metropolitan levels around a local purchasing policy, staff training, efforts to combat food waste and support for traditional cuisine. This shared vision allows the various political, institutional and academic actors to co-construct a "five-year action plan, with priorities based on the constraints and the meanings given to the actions. For example, raising awareness among parents in 2015 would not have been credible given the situation at that time" (Luc Lignon, director of food policy).

Since 2016, the MCA program has led to 25 actions which stem from the following four axes: a sustainable food purchasing policy, improvement of production management, optimization and modernization of meal distribution and raising awareness of sustainable food.

To assess this new school catering strategy, the DPA relies on the Urbal program, which was developed by a multidisciplinary team of researchers specializing in food issues. The main aim of this collaboration is to assist the DPA in its strategic thinking and decision-making by evaluating the impacts of food innovations in terms of sustainability in a multidimensional approach. Intoppa and Valette (2022) define the Urbal method as an assessment based on a qualitative and participatory approach, actively involving various stakeholders, such as suppliers, agents, children, parents and DPA. These participatory workshops allow exploration of the changes – expected or unexpected, positive or negative, real or potential – brought about by food innovations in terms of sustainability, while identifying the factors that may promote or hinder these developments.

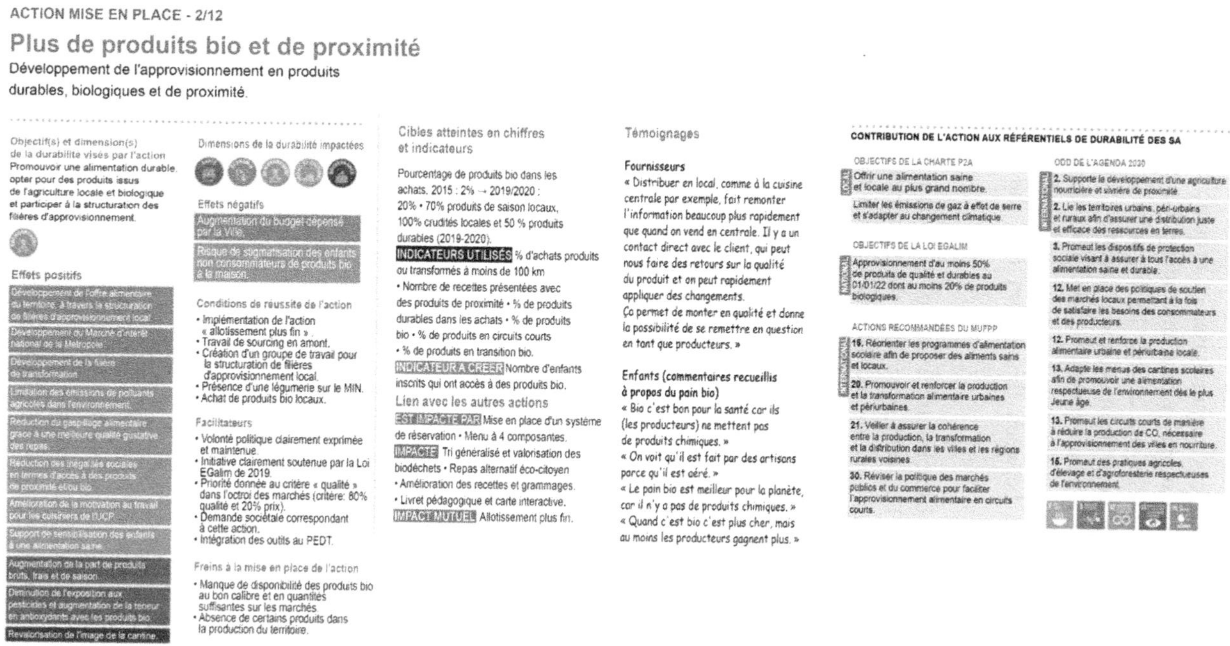

Figure 8.2. *Extract from the Urbal evaluation poster for the Ma cantine autrement program regarding the "More organic and local products" action – purchasing policy axis. For a color version of this figure, see www.iste.co.uk/ngoala/smartcity.zip*

Each action is evaluated in terms of its positive and negative effects on health, the economy, the environment, the social dimension and culture, as well as governance. For example, increasing purchases of organic and local products leads to developing the area's food supply while enhancing the canteen's image. On the other hand, children who have no access to organic food at home may feel stigmatized and question, or even devalue, their parents' food choices. Factors that aim to support and strengthen the action studied are also identified. For example, the Egalim law of 2019, as well as the trend at the societal level in favor of organic food, shows the coherence between this action and the more macro levels of analysis.

On the other hand, the obstacles are a lack of sufficient supply at the territorial level, that is, a local level. To be in sync with the systemic approach of the MCA program, the results of the action are linked to the other actions of the program. For example, the development of a more organic and local offering can be integrated into the awareness-raising axis through the educational booklet. Testimonies are also promoted, demonstrating the (at least stated) desire to integrate stakeholders. Finally, the impacts are put into perspective in terms of the benchmarks considered in the MCA program at different levels: the agroecological and food policy (P2A) defined in 2015 at the metropolitan level, the Egalim law at the national level and the Milan Pact at the international level.

Data collection and integration of the different stakeholders are conducted throughout the process of implementing the MCA program, from its design to its operationalization. Furthermore, a study of the impacts of the actions implemented, performed using a multidimensional approach to sustainability, provides an integrated view of the results.

The MCA program is now considered to be a success. Communication on this approach began "in 2018–2019, when they [political actors] began to understand the extent of the project's influence both in professional networks and also with international recognition through the Milan Pact, through requests for partnership and collaborative work" (Luc Lignon, director of the PA).

This experiment, carried out among children in the context of school catering, was conducted with an audience that was familiar and reachable for the DPA and the metropolis, which have competences for this audience. Accessibility to healthy food has objectively improved. It would nonetheless be interesting to study the impact of this program on perceived accessibility, both for children and parents. A recent study by Valette et al. (2022) on this specific case shows, for example, that some students' parents were concerned about the introduction of menus without foods of animal origin in school canteens. They feared this would be driven by cost savings, at the expense of users, and that it would deprive children of animal protein, which is appreciated and sometimes scarce in poorer households.

Skepticism and resistance to innovations aimed at promoting healthier eating underscore the importance of pedagogy in the context of school catering and demonstrate the challenges of objective accessibility. In addition to creating a healthy food environment, one essential objective is to encourage students to support and adopt the new measures, which are nonetheless aligned with objectives of better health and of considering environmental and societal issues.

Based on this experiment carried out in situ, the challenge for the city is now to deploy a sustainable food strategy aimed at all of the city's inhabitants as part of an approach of accessibility for all.

8.3.3. *The Food City: a project to broaden the target audience of the city's actions*

La Cité de l'Alimentation (Food City) is the result of a collective reflection initiated in 2019 by the DPA after the historic central kitchen exceeded its maximum production threshold. The project for a new central kitchen was approved in 2019 by the municipal council, with the aim of opening new premises in 2023. However, the election of a new team at the town hall in 2020 led to a rethink of this project, despite maximum production capacity having been reached.

Diagnostics of the central kitchen and design studies were repeated with the various elected stakeholders, technicians, academics, users, etc. A more ambitious project than just the creation of a new production unit was designed, to go beyond the target audience of children: "Food City is an augmented version of 'Ma cantine autrement'. [...] Children were at the center of 'Ma cantine autrement', and they are still at the center of the system, but reaching children is not enough, and so, if you like, Food City adds an additional line of work, an additional objective around solidarity food, around the solidarity and awareness aspect" (Luc Lignon, director of the PA).

These comments reveal the city's ambition to develop third spaces for food, open these collective kitchens to other audiences and create vegetable gardens. An educational bus will also be used to travel around the city, with the initial aim of raising awareness among schools, but the targeted audience may be broadened. "It means expanding the sustainable and supportive food system set up via 'Ma cantine autrement' for children in primary schools to all citizens, with particular attention to those in precarious situations" (report from city meeting).

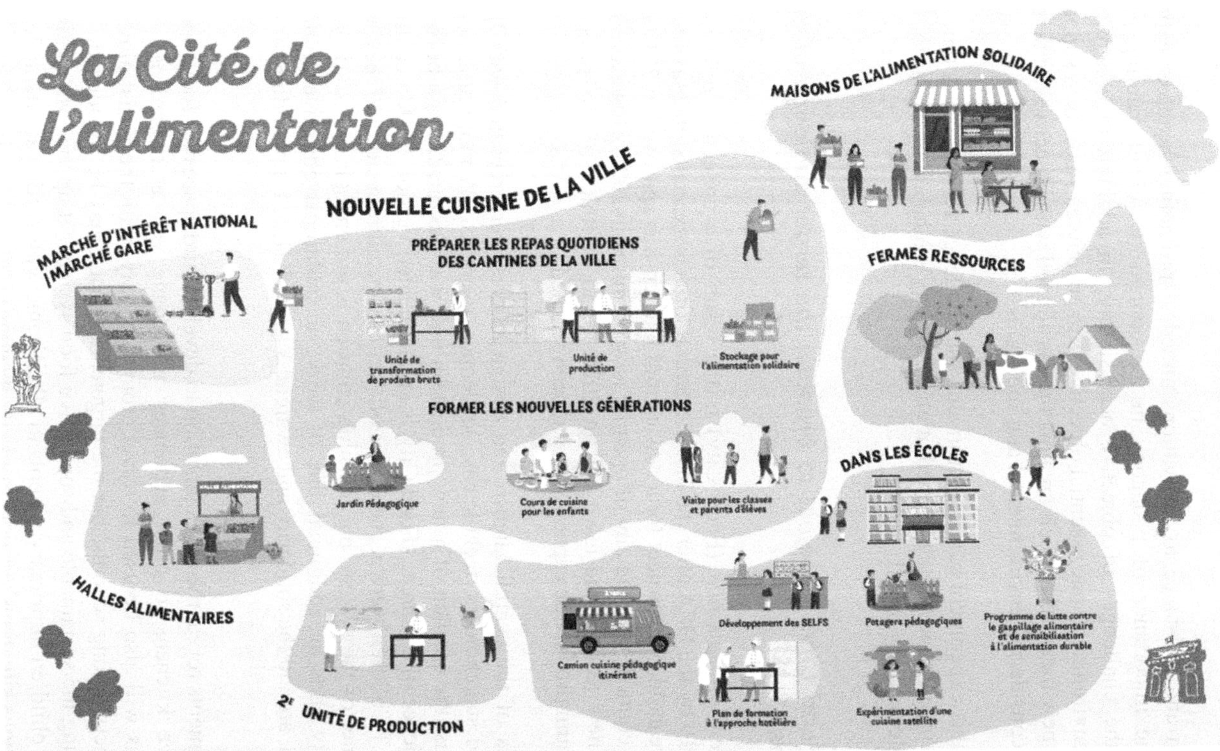

Figure 8.3. *Food City infographic (source: https://www.montpellier.fr/4826-le-projet-de-cite-de-l-alimentation.htm). The infographic covers food in schools, farms, production units and covered markets. For a color version of this figure, see www.iste.co.uk/ngoala/smartcity.zip*

The city's vision with Food City is to "sweep through the entire food system. We are re-examining it; we are trying to develop actions relating to sustainable development, but also to develop and unite as many people as possible around this idea of solidarity food" (Luc Lignon, director of the PA).

Looking at the Food City infographic (Figure 8.3), we can see that the different stages of the process, from production to consumption, are covered. Resource farms represent the place of production and the agricultural dimension, the wholesale market and the food halls constitute places of transport and supply, the central kitchen constitutes the tool of transformation and production, and the schools and the Solidarity Food Houses are spaces dedicated to consumption and sharing. The combination of these places provides support for teaching about sustainable food.

In the end, "the central kitchen is just one element of the final concept. In fact, the Food City's aim is to shine forth and spread throughout the city and the metropolis for all its inhabitants" (Luc Lignon, director of the PA). Food City operates in a larger food environment than that of the MCA program and therefore challenges the project's ability to respond to different audiences.

8.3.4. *La Maison de l'alimentation solidaire (Solidarity Food House): third place and target audience*

The Food City is, in a way, an assemblage of various concrete actions around food in the city. The Solidarity Food Houses are the first step in this widening of the target audience:

> We will reach audiences that we do not reach yet, residents in general. At this point, they are affected because either they have children who are in school, or they have family whose children are in school. But, on the other hand, they will be affected by the Solidarity Food Houses as soon as we start (Luc Lignon, director of the PA).

More precisely, the current project is to establish a Solidarity Food House (MAS) in the Celleneuve district. This MAS (provisional name, because voting for the name of this third place is in progress) aims to establish food as a common good. The atmosphere would be that of a popular canteen with accessibility to the greatest number, shared governance, but with values clearly defined from the start. This is not a form of social assistance, but a community project promoting diversity through food. Students could be vectors of this diversity, and it would therefore be possible to see this place as a laboratory of diversity.

In this context, the objective of the city's elected officials is to lead workshops presenting the project, followed by discussions in the form of brainstorming in the spirit of co-creating the place with those who may frequent it.

> Our idea is to have a solidarity food house for each political district, and there, to start with, the Solidarity Food House will be organized [...]. Governance will be organized in consultation with citizens. It is citizens, the local associations and the citizens of the neighborhood who will imagine and implement the Solidarity Food House; we will do that with them [....]. So they will necessarily be involved in the Food City. But it will happen progressively, step by step, in the neighborhoods, with the Solidarity Food Houses (Luc Lignon, director of the PA).

In this sense, the strategy of the MAS concept is interesting to study, because this place aims to bring together different audiences, who are not used to mingling, around initiatives relating to food. Ideally, stakeholders in the city's food policy want this third place to comprise of various living spaces. This would make it possible to buy healthy, local products, but also to cook and to eat on site. Activities around food are also envisaged (see Figure 8.4).

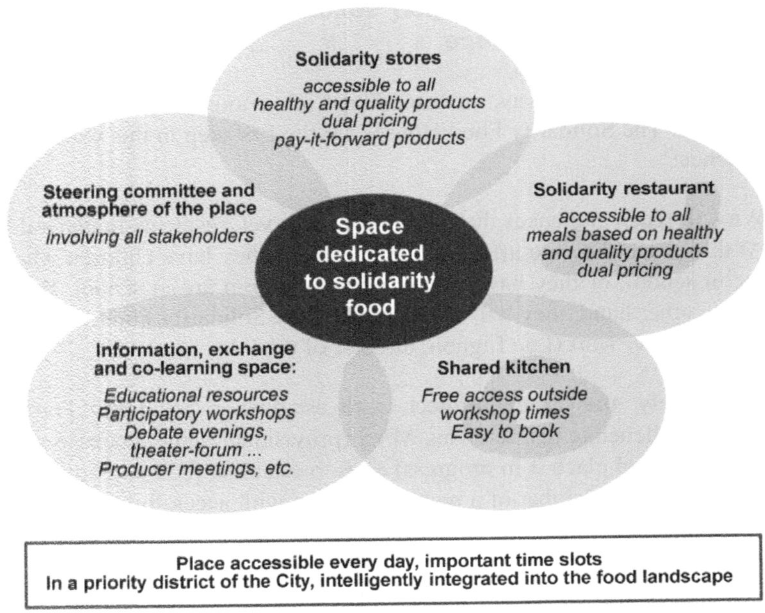

Figure 8.4. *Content of the Montpellier Solidarity Food House project (source: http://www.montpellier-alimentation-solidaire.xyz/?ComptesRendus)*

STEP 3: What are the must-haves (priorities)? What don't we want?

Must-haves	
Open space for meeting, free access Reception area Open to everyone (origins, social classes, age, etc.) Family place Open all the time Open to others' ideas A place to listen Listening, accepting difference, a "warm welcome" in the way of responding to people: first we welcome and listen, we take the time to discuss, etc.	**DIVERSITY!** **Caring welcome for all** **Openness to others** **Listening** ⇨ **Do not target only certain people** ⇨ **Think about how to implement social diversity: not easy!** ⇨ **Think about what will make the place hospitable for all social categories** ⇨ Draw inspiration from the training courses created by the Directorate General for Social Cohesion with ATD Fourth World (welcome, dignity, participation)
What we don't want	
A place stigmatized as "hipster or homeless"	**A stigmatized/excluding place for certain categories of person**

Figure 8.5. *Translated extract from expectations relating to MAS (source: http://www.montpellier-alimentation-solidaire.xyz/?ComptesRendus)*

The interest of this project for the City is that it allows the City to implement its self-declared mission to make healthy food accessible to all. However, in general, without expressing it explicitly, this type of third place mainly targets people in situations of food insecurity, those in need of support, while insisting on social diversity (Scherer 2022). Moreover, as far as the city of Montpellier is concerned, this question of the target audience is constantly raised. Indeed, those involved in the city's food policy are particularly aware of the difficulty of bringing together all types of publics in the same third place of food, and thus of pursuing the imperative of social diversity that they constantly mention. The risk associated with this type of place is its "hipster or homeless" connotation (report from city meeting).

However, the imperative of social diversity raises questions, because it implies increasing the number of target audiences to achieve diversity. However, each target audience has different expectations regarding food. Thus, the often-mentioned dual pricing scheme relates to both individuals in precarious situations and people wishing to show solidarity; but can we consider this to be diversity?

This also leads the city to wonder about the expectations of people who encounter food difficulties, but whom they do not know well. Such is the case for students.

Indeed, since the Covid-19 crisis, the implementation of one-euro meals by CROUS (Regional Center for University and School Operations) has not been enough to meet the need for accessibility to healthy quality food for students in precarious situations. Students, for whom food represents an average of 20% of the budget, are in fact identified as a group particularly exposed to food insecurity (Nikolaus et al. 2020; Beacom et al. 2021), defined as limited access to adequate food due to lack of money or other resources (Gundersen and Ziliak 2015), as well as food precarity (Paturel 2019).

It is with the desire to better understand them, as well as determine their expectations about the future MAS, that brainstorming workshops were held in Montpellier in January 2023.

The students described several places combined into one, with individual/collective or leisure/work type spaces, whose ambiances should be made distinct to respond to everyone's needs. Training, in different formats (workshops, conferences, courses, etc.), is an important service for students. The MAS is seen as a place to learn about nutrition.

However, it is interesting to note the points of divergence between the MAS as seen by students and the project initially designed by the city.

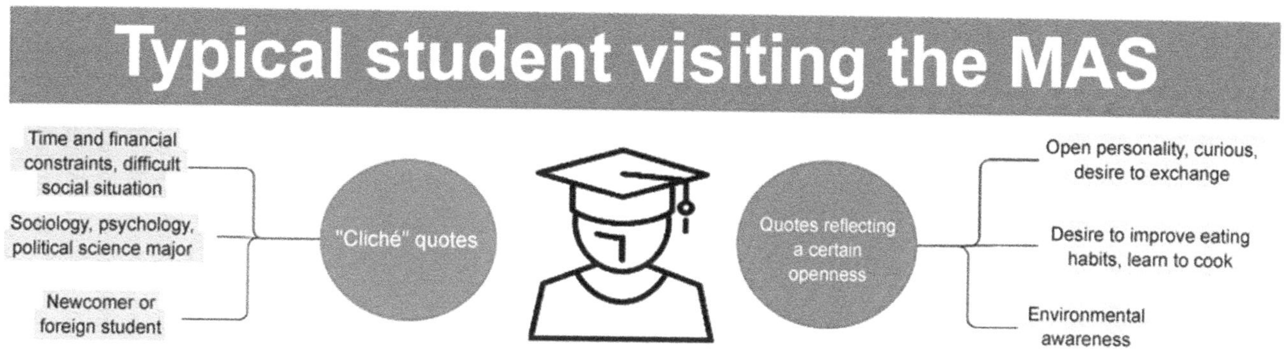

Figure 8.6. *Result of brainstorming session with students about MAS. For a color version of this figure, see www.iste.co.uk/ngoala/smartcity.zip*

The universal character/social mix is debated. Although students all agree with the principle of diversity, the description of their ideal MAS sometimes seems to contradict this principle. Some comments reflect a desire to take ownership of this place to the point of almost transforming it into a place dedicated to students. The ideal MAS would thus be located on a university campus. It would give students the right to benefits such as ECTS academic credits in exchange for volunteering (for example, by serving meals), discounts on presentation of a student card, or even an opportunity to obtain student jobs, internships, or work–study programs.

Furthermore, students who discuss communication around this project perceive it as digital above all. The following actions were proposed: an Instagram presence and creation of engaged publications, TikTok, Twitter, creation of a YouTube channel with regular uploads, nutrition/health blog, website and sending out mailings within the university. While some participants explained that these ideas were solely intended to attract the student target audience, this suggests that the use of many different communication tools and messages will be necessary to adapt to each target.

Other students describe a fairly precise profile of the type of student who would frequent the MAS. Accordingly, interest would vary depending on the type of student.

The purely food-related character of the initiative is also questioned. While students do not question the value of offering activities around food, they would prefer more diversity in the topics covered. They thus perceive the MAS as a lively and vibrant place thanks to regular events, not necessarily directly linked to food (music, culture, exhibitions, permanent gallery, art, garage sales, bingo, activities and games made available). For them, a friendly atmosphere is reflected in the presence of entertainment (board games, table football, concerts, cinema, etc.).

8.4. Conclusion

"This capacity to contribute to the transformation of food systems is in fact weak if they do not have the potential to be replicated, imitated, amplified, supported, disseminated, linked together" (Valette et al. 2022). This chapter has illustrated these remarks by confronting theoretical concepts relating to the "smart" food environment with the challenges of cities trying to take ownership of the issue of food and food sovereignty. Using the example of the city of Montpellier, we have shown that acting for a smart urban food environment involves long-term projects, including as many stakeholders as possible. While some of the city's programs have already been implemented, notably "Ma cantine autrement", discussions are still underway to transform actions that were initially unrelated to each other into a real network, a food city. The objective is to achieve accessibility to healthy and quality food for all residents. However, this chapter has revealed the difficulties of achieving this objective for a city, through the questions raised by the Montpellier projects. So, to answer the initial question, "how can we make the food environment smart in a city and help improve residents' access to healthy food?", it is possible to reason in several stages, as the city of Montpellier has done:

– Bring together stakeholders around a project aimed at a well-known audience. This allows the foundation to be established and the issues to be explained to the various stakeholders who need to be involved in questions of food.

– Expand the network and include new initiatives to draw up a roadmap that leaves no target audience behind.

– Implement the actions described one by one, notably through the creation of third places, taking care to understand the needs of each target audience, in order to limit the risks of stigmatization or connotation of a third place (brainstorming sessions for each target audience, especially with audiences little known to the city).

8.5. References

Albino, V., Berardi, U., Dangelico, R.M. (2015). Smart cities: Definitions, dimensions, performance, and initiatives. *Journal of Urban Technology*, 22(1), 3–21.

Beacom, E., Furey, S., Hollywood, L.E., Humphreys, P. (2021). Stakeholder-informed considerations for a food poverty definition. *British Food Journal*, 123(2), 441–454.

Bernardin, S. and Jeannot, G. (2019). La ville intelligente sans les villes : interopérabilité, ouvertures et maîtrise des données publiques au sein des administrations municipales. *Réseaux*, 218, 9–37.

Calise, T.V., Chow, W., Ryder, A., Wingerter, C. (2019). Food access and its relationship to perceived walkability, safety, and social cohesion. *Health Promotion Practice*, 20(6), 858–867.

Cooksey-Stowers, K., Schwartz, M.B., Brownell, K.D. (2017). Food swamps predict obesity rates better than food deserts in the United States. *International Journal of Environmental Research and Public Health*, 14(11), 1366.

Daviron, B., Perrin, C., Soulard, C., Menant, F. (2017). Histoire des politiques alimentaires urbaines en Europe, de la ville antique à la ville industrielle. In *Construire des politiques alimentaires urbaines. Concepts et démarches*, Brand, C., Bricas, N., Conaré, D., Daviron, B., Debru, J., Michel, L., Soulard, C. (eds). Éditions Quæ, Paris.

Dey, S., Saha, S., Singh, A.K., McDonald-Maier, K. (2022). SmartNoshWaste: Using blockchain, machine learning, cloud computing and QR code to reduce food waste in decentralized web 3.0 enabled smart cities. *Smart Cities*, 5(1), 162–176.

Dupuy, G., Minster, C., Watel, R. (2011). Environnement urbain et obésité : peut-on transposer en Europe les approches nord-américaines ? *Annales de géographie*, 682, 604–628.

Eveno, E. (2018). La Ville intelligente : objet au cœur de nombreuses controverses. *Quaderni*, 96, 29–41.

Fattibene, D., Recanati, F., Dembska, K., Antonell, M. (2020). Urban food waste: A framework to analyse policies and initiatives. *Resources*, 9, 99.

Goldstein, B., Birkved, M., Fernández, J., Hauschild, M. (2017). Surveying the environmental footprint of urban food consumption. *Journal of Industrial Ecology*, 21(1), 151–165.

Gordon, R., Carrigan, M., Hastings, G. (2011). A framework for sustainable marketing. *Marketing Theory*, 11(2), 143–163. doi: 10.1177/1470593111403218.

Gundersen, C. and Ziliak, J.P. (2015). Food insecurity and health outcomes. *Health Affairs*, 34(11), 1830–1839.

Hamilton, R., Thompson, D., Bone, S., Chaplin, L.N., Griskevicius, V., Goldsmith, K., Hill, R., John, D.R., Mittal, C., O'Guinn, T. et al. (2019). The effects of scarcity on consumer decision journeys. *Journal of the Academy Marketing Science*, 47, 532–550.

Hasnaoui Amri, N., Michel, L., Soulard, C.T. (2020). Une politique alimentaire à Montpellier. La transition agroécologique vecteur de compromis politique. In *Quand l'alimentation se fait politique(s)*, Fouilleux, È. and Michel, L. (eds). Presses Universitaires de Rennes, Rennes.

He, S.Y., Lisa, C., Bhada-Tata, P., Van Woerden, F. (2018). What a waste 2.0: A global snapshot of solid waste management to 2050. Report, Urban Development.

Intoppa, B. and Valette, E. (2022). Approches collaboratives pour l'identification d'indicateurs d'évaluation : usage de la démarche URBAL. Report, Chaire UNESCO Alimentations du monde, Montpellier.

Kakaei, H., Nourmoradi, H., Bakhtiyari, S., Jalilian, M., Mirzaei, A. (2022). Effect of COVID-19 on food security, hunger, and food crisis. In *COVID-19 in the Sustainable Development Goals*, Dehghani, M.H., Roy, S., Karri, R.R. (eds). Elsevier, Amsterdam.

Kim, H.M., Sabri, S., Kent, A. (2021). Smart cities as a platform for technological and social innovation in productivity, sustainability, and liveability: A conceptual framework. In *Smart Cities for Technological and Social Innovation: Case Studies, Current Trends, and Future Steps*, 1st edition, Kim, H.M., Sabri, S., Kent, A. (eds). Academic Press, London.

Lake, A. and Townshend, T. (2006). Obesogenic environments: Exploring the built and food environments. *Journal of the Royal Society for the Promotion of Health*, 126(6), 262–267.

Maye, D. (2019). 'Smart food city': Conceptual relations between smart city planning, urban food systems and innovation theory. *City, Culture and Society*, 16, 18–24.

Michel, L. and Soulard, CT. (2019). Putting food on the regional policy agenda in Montpellier, France. In *Designing Urban Food Policies: Concepts and Approaches*, Brand, C., Bricas, N., Conaré, D. et al. (eds). Springer, Cham.

Nikolaus, C.J., An, R., Ellison, B., Nickols-Richardson, S.M. (2020). Food insecurity among college students in the United States: A scoping review. *Advances in Nutrition*, 11(2), 327–348.

Palma Lampreia dos Santos, M.J. (2016). Smart cities and urban areas – Aquaponics as innovative urban agriculture. *Urban Forestry & Urban Greening*, 20, 402–406.

Parekh, N., Ali, S.H., O'Connor, J., Tozan, Y., Jones, A.M., Capasso, A., Foreman, J., DiClemente, R.J. (2021). Food insecurity among households with children during the COVID-19 pandemic: Results from a study among social media users across the United States. *Nutrition Journal*, 20(73).

Paturel, D. (2019). Diagnostiquer la précarité alimentaire à une échelle locale. Report, Chaire UNESCO Alimentations du monde, Montpellier.

Scherer, P. (2022). Des tiers-lieux au service de nouvelles solidarités alimentaires. *Cahiers de l'action*, 58, 16–26.

Swinburn, B.A., Caterson, I., Seidell, J.C., James, W.P.T. (2004). Diet, nutrition and the prevention of excess weight gain and obesity. *Public Health Nutrition*, 7(1a), 123–146.

Testard-Viellard, P. (2022). La ville est-elle l'avenir de l'humanité ? *Carnets de Sciences – La revue du CNRS*, 13.

Valette, E., Lepiller, O., Bonomelli, V. (2022). Des innovations à la transition des systèmes alimentaires : comment penser les conditions et les modalités de leur changement d'échelle ? *Géocarrefour*, 96(3).

Vitman-Schorr, A., Ayalo, L., Khalaila, R. (2019). Perceived accessibility to services and sites among Israeli older adults. *Journal of Applied Gerontology*, 38(1), 112–136.

Wang, D., Brown, G., Liu, Y. (2015). The physical and non-physical factors that influence perceived access to urban parks. *Landscape and Urban Planning*, 133, 53–66.

Yamaguchi, M., Takahashi, K., Hanazato, M., Suzuki, N., Kondo, K., Kondo, N. (2019). Comparison of objective and perceived access to food stores associated with intake frequencies of vegetables/fruits and meat/fish among community-dwelling older Japanese. *International Journal of Environmental Research and Public Health*, 16(5), 772.

9

Smart Culture, Immersion, Co-creation and Public Engagement

9.1. Introduction

If culture is at the heart of the societal project of public policy, its democratization is the subject of an important ideological debate. On one side of this debate are those who wish to promote a universal culture for all, which satisfies the expectations, constraints and interests of the public to make it accessible to the large number of people. On the other side are those who want a more elitist culture that encourages each citizen to improve and transcend themselves through erudite knowledge. For Pagès (2020), the challenge is to promote a cultural standard in the logic of creating a "cognitariat", aiming to "prioritize access to the arts and culture to train educated and informed citizens, but also curious, creative individuals open to the growing complexity of the world". These partisans fear that "instead of popularizing culture, we will vulgarize it" and campaign for limited distribution to an enlightened and deserving public (Godin 2011, p. 164). However, as noted by Francis Lacloche, advisor to the Ministry of Culture and Communication from 2009 to 2012, the "real obstacle to a policy of cultural democratization is culture itself. A certain idea of culture, widespread in diverse parts of society, leads, under the guise of expectation and excellence, to a process of social intimidation" (Godin 2011), making this second approach a dead end.

Smart culture – or the integration of technologies of cultural mediation into the smart city to encourage access to culture – is the bearer of many promises to reconcile this opposition between populism and elitism, but also risks. In the context of museums, Guintcheva and Lagier (2019) notably distinguish a wide variety of technological devices focused on content and individual learning

Chapter written by Olga GONCALVES, Camille LACAN and Gilles N'GOALA.

(websites, video guides, online collections, digital works, touchscreens, terminals) or on interaction and sharing with family and friends (personalized visitor trails, gamification, reception robots, visitor-produced content). They also highlight the expansion of immersive systems in museums (3D, 360°, holograms, etc.). Each technological system certainly raises specific issues in the museum context, as in other artistic, cultural or heritage fields (live shows, architecture, monuments, historical and natural heritage, etc.). Not all technological systems respond to the same challenges, produce the same experiences, or lead to the same problems of adoption and appropriation. However, movement toward the digitalization of cultural offerings and experiences is massive, protean and inevitable throughout the field of culture. The digital refers both to new digital media and to a more intensive use of data collected about works and audiences. This growth is encouraged by authorities and cultural institutions as well as by territories and destinations seeking to develop an attractive, innovative and varied artistic and cultural program.

By recognizing the diversity of audiences and the role of personal experiences online or offline, the rise of a smart culture thus disturbs the cultural field, raising fears of a deterioration in knowledge and pleasure in the relationship with works; this could counter an elitism for all with a leveling for everyone. Several questions are raised by the technological developments affecting the cultural, artistic and heritage fields.

– Does the digital transition, operating throughout society, constitute a threat or an opportunity to legitimize actors in the cultural field? What are its effects on the strategies, resources and organizations of cultural stakeholders?

– Is cultural experience enhanced or devalued by integrating new technological media? What are their contributions in terms of transmission of knowledge, pleasure, play, self-esteem, etc.?

– Is increased access to culture a factor in the public's appropriation of cultural experience?

– Since cultural experience is the result of co-creation between an audience, a work, and a particular context, are audiences more engaged or disengaged when they use these new cultural media?

– Is smart culture synonymous with smart people or does it lead to new inequalities and divisions within society? Is the digital divide another cultural divide?

9.2. The digital in search of legitimation

The integration of digital technology into cultural institutions has been imposed by a double imperative – organizational to improve its functioning, but also societal, due to new public aspirations (Alexandre et al. 2022). The adoption of digital devices in the cultural field responds to a growing need for effectiveness and efficiency in cultural organizations, to the feeling that the digital transition makes changes in practices and strategies inevitable; and also due to the conviction of certain stakeholders that technology is not necessarily contradictory with the values and standards of the world of art and culture. Although digital technology still provokes resistance and controversy, discourses legitimizing the place of digital technology in culture are gradually becoming established.

The contributions of digital technology in management terms are indeed numerous, since digital technology facilitates the organization, planning, implementation and control of cultural activities. For managing a catalog of works in a museum, exhibition, purchase or even their conservation, digital tools offer a useful function as a support for cultural activity (Rivière and Bourliataux-Lajoinie 2017). Information is archived, prioritized and connected with partners through the information system. For example, the Culture Pass offers young people the opportunity to access many cultural places using an app that covers the whole cultural industry. These tools also allow cultural organizations to adopt practices from commercial organizations: customer relationship management, demand flow analysis, real-time demand management, online ticketing and targeted communication, among many others. In the quest to optimize and reduce operating costs, digital technology improves support functions (see Box 9.1).

At the societal level, culture could hardly ignore the new expectations of audiences who aspire to a need for connection and to digitalization of their experiences (Alexandre et al. 2022). Digital technology has become a vector of empowerment for the public, giving them control over their environment. Apps such as Affluences have seen great success in allowing users to find out, for example, museum opening times and waiting times. Thus, the *bring your own device* trend has entered cultural spaces, making cultural places a space like any other, where digital technology is present among visitors (Jarrier and Bougeon-Renault 2020). In short, it is difficult for cultural venues to sidestep the digital model when it is omnipresent elsewhere.

However, at the same time, digital technology raises a certain number of reservations on the part of practitioners who are aware of the frictions emerging from it (Pagès 2020). First, there is a debate about the meaning and usefulness of digital technology, as the sudden arrival of digital technology in traditionally conservative worlds has raised many questions. As Sandri (2016) reports, "hosts are divided, because they see that digital is more of an add-on than a necessity. We have the content, the host could do it without digital support". Between a fad and real utility, cultural professionals are questioning the true utility of digital technology.

Digital devices at the Fabre Museum

By Anne Le Cabec, digital project manager, Fabre Museum, Montpellier.

For almost a decade, the Fabre Museum has used digital mediation technologies to support and enhance its collections and exhibitions. Using an active multimedia service that draws on its remarkable collections, it has established itself as a testing ground for innovative initiatives, on site and outside.

Development and evaluation of a range of varied devices now allows us to lay the basis for a rational digital deployment strategy at a scale sufficient for our collections and over the long term.

Development of these devices is always preceded by a reflection allowing evaluation of the added value of digital technology compared to other types of mediation. The challenge in fact is to use these technologies to promote encounters with works and to invent new ways of sharing knowledge and experiences.

From downloadable mobile apps to online apps

The visitor assistance apps offered for our major summer exhibitions between 2016 and 2018 met with moderate success, explained largely by constraints related to downloading (need for space on the phone and a good network).

Since 2020, we have preferred content accessible online, on the museum website, to downloadable applications. Thus, each exhibition is associated with a web interface which combines text/image presentations, audio guides and interactive experiences where appropriate (e.g. the game *Collector*, designed in parallel with the exhibition *Sharing Beauty – 15 Years of Acquisitions at the Fabre Museum* in 2022).

Fixed devices, relating to the works

Since 2013 and still today, we have been using our touchpads to enhance visits to our summer exhibitions and make the visitor an actor in their visit experience, following three objectives.

Unveiling the invisible

These interactive experiences allow us to unveil the scientific imagery of certain works (X-rays, infrared, ultraviolet, etc.) and reveal the artist's creative process (underlying compositions and preparatory drawings).

They also allow restoration of historical contexts (Frédéric Bazille, *The Youth of Impressionism* (2016), Jean Ranc, *A Montpellier Resident at the Court of Kings* (2020)).

They are also used, in certain exhibitions, to give visitors the opportunity to "flip through" books or sketchbooks in their entirety.

These generally meet with great success (Frédéric Bazille, *The Youth of Impressionism* (2016): 5,195 consultations; *Picasso. Donner à voir* (2018): 10,744 consultations).

Making the visitor an actor in the experience

These interactive devices give visitors the pleasure of learning for themselves. They are no longer passive, but when put into a situation, they engage in the experience to better appropriate the knowledge. This is evidenced by the interface developed for the exhibition *Dans le Secret des works d'art* (2018) and now adapted for the web, the public and schoolchildren. This experience allows the visitor to play the role of the scientist–restorer who observes the material of the work and discovers the different scientific images.

Restoring the works to their creative context

The latest 3D modeling technologies have enabled the reconstruction of the prestigious Aeneas Gallery at the Palais-Royal in Paris, restoring the history of three monumental paintings by Antoine Coypel, emerging from the decor that adorned this gallery throughout the 18th century. Visitors can navigate this gallery in 3D using a touchscreen set up opposite the works, in the museum's column gallery.

Projections and VR: immersive experiences to enrich discovery of an artistic process

Immersive experiences were offered at the end of the trail for the last two summer exhibitions to immerse the visitor in a creative journey (*Louis Gauffier's Trip to Italy* in 2022) and technical experiments (*Germaine Richier, a Retrospective* in 2023).

Immersive projections have been very successful, as they allow for a collective sensory experience and appeal to visitors' imagination and sensitivity.

The VR experience, deployed for the Louis Gauffier exhibition, was also very popular with the public and could be repeated for an appropriate subject, despite the logistical constraints.

Box 9.1. *Digital devices at the Fabre Museum (Anne Le Cabec, digital project manager, Fabre Museum)*

Reluctance also stems from a risk of losing control over the cultural object. Cultural professionals present themselves as the guardians of culture. They have the mission of making culture accessible to all, but also, and even primarily, of preserving and protecting it. However, audiences share their experiences on their social networks, whether by posting photos or through vlogs – videos through which each Internet user can then visit the museum from home. This practice is a double-edged sword. Indeed, if it allows expanding the museum exhibition's audience, it also raises the question of decreasing physical attendance at places and their uses in digital spaces, which are open and out of control of the world of culture.

These reluctances are often based on certain experts' fear of a loss of legitimacy. For example, mediators possess expertise and scholarly knowledge about culture. This cultural capital confers legitimacy upon them to present, decipher and criticize works in this field. If this capital is transferred to digital services, accessible to all, the role, identity and legitimacy of cultural mediators will have to be redefined. Another fear of the experts is also to see the knowledge and the level of demand required for access to works decline in favor of a culture of consumption and *edutainment*, which would place entertainment on the same level as public education.

The objective of cultural enrichment for all collides here with the logic of creating value for everyone. Bourdieu's model of "cultural legitimacy" emphasized the extent to which culture responded to both a need for "social distinction" and submission to the norms of a group of belonging (in particular, codes learned within certain social classes). Cultural mediation thus sought to bridge these social inequalities by disseminating knowledge to all.

However, in value creation strategies, individuals are not solely determined by their social environment; they deliberately decide and choose their cultural consumption based on the value it brings to them. From this perspective, therefore, one of the keys to the transmission of culture resides in enrichment of experience.

9.3. Toward new cultural trails: creation or destruction of value for audiences

Combining the challenge of education with that of aesthetic pleasure and play, digital technology represents an important lever for the smart city in creating cultural value. The cultural field is particularly conducive to these experiential strategies, since cultural consumption is more associated with the notion of aesthetic pleasure and intrinsic motivations; a work of art is appreciated for itself and not for its utilitarian functions alone.

Thus, for cultural institutions and public authorities, technological devices respond to new challenges, such as developing a richer experience focused on social, hedonic and ludic dimensions.

From the experiential perspective (Holbrook and Hirschman 1982), traditional cognitivist models give way to more emotional approaches in which the public is looking for experiences, among which culture becomes an object of self-satisfaction and self-realization (Maffesoli 1990). Depending on what the offer represents for them, they will have a holistic perception of the culture (Bourgeon and Filser 1995). Thus, experiential culture can be defined as "a subjective state of consciousness accompanied by a variety of symbolic meanings, hedonistic responses, and aesthetic criteria" (Holbrook and Hirschman 1982, p. 132). It is about going beyond the purely functional or utilitarian vision of a product/service to assert hedonistic gratification in a social context, since culture is perceived as a source of sensations, entertainment, pleasure and emotions, going as far as expressing the public's quest for identity (Cova and Cova 2001). Culture can thus be understood in two different ways: a functional/utilitarian view, focused on quality of service, practicality and knowledge (epistemic value), and an experiential view which values aesthetic pleasure, play, emotions, escape from everyday life and symbolic value (social status, self-esteem, etc.) in the understanding of experience (Pieters and Van Raaij 1988).

Through approaches combining physical and digital media in a "phygitalization" movement, cultural institutions are now seeking to offer audiences experiences enriched by different types of value (other than knowledge), differentiated, immersive and interactive, allowing awareness and access to art and culture for the greatest number (Jarrier and Bourgeon-Renault 2020). This trend is illustrated by the creation of "off-site" experiences, the use of digital technologies to gamify artistic and cultural experiences, the establishment of creative and participatory workshops or the dissemination of immersive technologies such as virtual reality or augmented reality. Digital technology has thus become essential along cultural trails in the hope of better promoting works and ensuring a smooth and optimal experience, whether before, during or after the cultural experience.

Creation of value through technological devices occurs at the three stages of the cultural experience.

Before the experience, digital devices are first and foremost a vector of interaction for cultural institutions and destinations with potential visitors. Websites and mobile applications (m-services) help provide reliable, contextualized and up-to-date information on the agenda of artistic and cultural activities (Guintcheva and Lagier 2019). Their interest is important for meeting the expectations of visitors in the information search phase (Rivière and Bourliataux-Lajoinie 2017) and contributes to making the visitor a smart shopper of art and culture through detailed knowledge of the most relevant activities in terms of their individual objectives (having fun, learning, creating social ties, etc.). In a logic of proactivity, digital devices aim to attract visitors through an attractive presentation of activities. This is the case, for example, for destination management organizations, which set out their offer on digital networks using branded content strategies (Uşakli et al. 2017). The presence of touch terminals and other interactive display tools within physical spaces creates real added value in welcoming visitors and personalizing their experience (Jarrier et al. 2019). These devices thus facilitate, before the experience, not only the dissemination of information, but also its presentation to different audiences.

During the experience, cultural mediation devices play a predominant role in shaping visitor experience. Whether they are owned by the public (smartphones, connected objects) or made available by a cultural or artistic institution (augmented or virtual reality headsets), the challenge of these devices is to provide a richer cultural experience by stimulating the senses, reasoning and imagination; in other words, to reconcile public awareness/education about art and culture with a more ludic and recreational dimension (Jarrier et al. 2019). These cultural mediation devices are of diverse natures and range from simple visitor guidance (audiovisual guides; smartphone applications), which provides the visitor with a virtual presence (Lemoine and Salvadore 2018; Guintcheva and Lagier 2019), to devices proposing to augment or even to transport the visitor into a hyper-reality (Graillot 2005). However, augmented reality refers to a "juxtaposition of a layer of digital information with a physical representation of reality" (Lachance and Bruna 2019); hyperreality leads to the creation of a new virtual environment within which the individual can interact and therefore implies an additional experience for the visitor (Bec et al. 2019).

Cultural mediation devices during the experience can be classified according to the degree of immersion generated by the device for the user and according to the possibility of interaction to control the experience (Carrozzino and Bergamasco 2010). Through immersion, the device aims to bring the visitor into a state of *flow*: a psychological state of cognitive absorption where the individual experiences "an exceptional (extraordinary and unforgettable) moment during which what we feel, what we wish, and what we think are in total harmony" (Csikszentmihalyi 1997, p. 29, in Carù and Cova 2006). The interest here is in reducing the effect of weariness that intervenes in intellectual activities (visiting a museum), particularly among children (Guintcheva and Lagier 2019). Through interaction, the device aims at the visitor gaining a better grasp of the cultural object. The individual is no longer passive in the experience but becomes a full-fledged actor. The resulting experience is thus co-created by the individual and the cultural object.

After the experience, digital devices create social and community value. For cultural institutions, digital technology makes it possible to develop relationship marketing, allowing them to maintain a link with their visitors through brand pages on social networks and online communities (Uşakli et al. 2017). Digital technology thus extends the educational and entertainment mission of cultural institutions into these digital social spaces. Digital devices can therefore facilitate visitors' connections with the cultural institution as well as among themselves, for example, by encouraging them to share their lived experience on social networks (photos on Instagram), and thus trigger imitation effects among new visitors.

In short, digital cultural mediation devices can create value before, during and after the cultural experience. However, some devices do not always enrich the experience for different audiences (Rivière and Bourliataux-Lajoinie 2017) or are they fully accepted or used (Lemoine and Salvadore 2018). Digital technology can only create value if audiences and users actively participate in co-creating the experience and simultaneously embrace these technological and cultural objects. However, the diversity of uses and users, who differ in their expectations and capacities regarding digital technology, does not guarantee creation of value for all, but on the contrary may destroy it by reducing the educational content previously transmitted by cultural mediators without providing the entertainment, pleasure, social connection and play that digital devices could produce. De Blas et al. (2015) thus show that interactive mediation tools developed by a museum convey positive values (freedom, generosity, public service, modernity and dynamism) and positive perceptions (pleasure), but are also associated with a loss of conviviality that may degrade the proximity of identity between the public and the art museum.

9.4. From accessibility to the appropriation of cultural experience by the public

Access to culture is a right for all citizens. Particular emphasis is placed on financial accessibility, which notably leads to a policy of general or targeted free access (young people, job seekers, etc.) and to physical accessibility, in particular for people with reduced mobility, vulnerable people, or those unable to travel. Digital technology thus allows financial barriers to be overcome – the devices are generally free – and allows everyone to access artworks remotely without fear of others' opinion. By reaching out to audiences without waiting for them to knock on the door of cultural venues, digital technology often aims to overcome barriers that seemed insurmountable to many audiences.

The recent acceleration in the development of digital devices is a response to both cyclical and structural factors.

First, the exceptional period of Covid-19 probably accelerated this movement toward the digitalization of experiences. Many heritage and cultural sites saw an increase in attendance in 2019. An example is the International City of Comics and Images in Angoulême, which, through partnership actions aimed at raising awareness among children, increased attendance by 6% between 2018 and 2019. However, the Covid-19 health crisis in 2020 (lockdowns, gradual reopenings subject to capacity constraints, vaccination passes, wearing masks, etc.) brutally disrupted the sector and showed the extent to which culture was an essential element in the lives of many French people (Müller and Schreiber 2022). The Covid-19 pandemic indeed highlighted how much the French were yearning for culture and hungry for digital technology during the phases of lockdown and then in the first stages of ending lockdown (closure and restriction of access to museums, bans on gathering in performance halls, etc.). Initiatives therefore proliferated to maintain links with audiences who were temporarily distanced or prevented from accessing cultural venues, like what many people experience in normal times. For example, the Fabre Museum in Montpellier offered games and virtual tours of exhibitions (Soulages, in particular) on Facebook and Instagram to raise awareness among its subscribers about its collections. The Muséum d'Angers broadcast a game, *Animal Crossing*, to the public wishing to visit the rooms for 30 minutes and ask questions of the archaeologists present. However, as Combaluzier (2020, p. 11) notes in her Master's thesis, one had to "subscribe to Nintendo Switch Online and to have access to Skype, many elements which distant audiences or audiences without computer tools such as the elderly, for example, cannot access". The same questions have arisen in other cultural universes, such as the theater, where the proliferation of filmed content or live broadcasts (for example, on streaming platforms) leads to questions about the nature and scope of the experience. During the lockdown, for example, a platform was set up to highlight the work of actors reduced to inactivity[1]. Thus, the pandemic period served as a *crash test* in the field of culture, which has sometimes been able to demonstrate resilience in a difficult context, but which has also often noted its profound fragility and unpreparedness in the face of the digital transformation and the habits of audiences that change with technological advances. The transformation of cities into "smart cities" is a game changer, as "art and culture can be considered a fundamental aim of smart city projects" (Meijer and Bolivar 2016).

Second, in a more structural way, the ambition of cultural stakeholders is to target new audiences who are far removed from culture through digital devices with which they are potentially more familiar (mobile applications, social networks, websites, etc.). In museums, for example, new strategies are being developed to attract "more novices, families, non-experts" in an objective of cultural democratization (Guintcheva and Lagier 2019). Indeed, in various cultural locations, digital devices (guided tours and online conferences, augmented reality, escape games, etc.) aim to facilitate the embrace of art, culture or even heritage by more or less experienced audiences. Digital technology, considered as a cultural mediation tool, is a "lever facilitating experience" (Guintcheva and Lagier 2019) and a means

1 See: https://theatre-confine-en-reseau.com/.

of improving the encounter between art/culture and "less experienced audiences, looking for experience and social connection" (Debenedetti et al. 2009). Thus, if it is asserted that, in the cultural and artistic domain, the physical encounter with the work constitutes an indisputable moment of truth, the challenge is more than ever to combine, complement and augment it using new digital devices.

Digital technology is often invoked, rightly or wrongly, as an instrument for combating inequalities in access to culture and for better social inclusion of remote or impoverished populations (young people, elderly people, inhabitants of isolated rural areas, etc.). It accounts better for the heterogeneity of audiences in terms of culture (expectations, constraints, tastes, uses, etc.) and offers varied and renewed means of access. However, some nuances must be added to these conjectures.

Digital technology does not completely remove the barriers to access of culture. For Laurent (2022), accessibility refers to the perceived ease for an individual of accessing an object, a place or an experience through their own capacities (cognitive, physical, financial, material, digital, etc.) in a given environment (real and virtual). Thus, the use of free-of-charge access and digital media does not prevent some audiences from encountering difficulties in access, notably because their cognitive (knowledge, education, etc.), physical (material, quality of connection, etc.), psychological (social identity, self-confidence), or digital (digital literacy) capacities may prove insufficient in regard to the proposed measures. In particular, the audiences affected by exclusion from the cultural field are sometimes the same as those affected by digital exclusion (disadvantaged populations, very elderly people, etc.). The digital transformation of cultural experiences may thus potentially reveal a double exclusion – digital and cultural – instead of promoting better inclusion and social cohesion.

Accessibility is thus a necessary, but not sufficient, condition for the appropriation of a work, a place or a cultural experience (Carù and Cova 2003). For Mifsud et al. (2019), appropriation refers to the process of "making suitable for a destination, adapting, conforming", which assumes that the user or patron can divert the object in question to adapt it to what they are (expectations, constraints, etc.), and to take possession – at least psychologically – of the object in question. Thus, audiences are called upon to develop themselves, familiarize themselves, develop expertise (mastery), adapt, create and take ownership of their cultural experience. However, cultural works, places, experiences and features remain difficult to "divert" and appropriate for a large part of the public. The primacy of cultural supply over public demand often limits this appropriation by the public, apart from those devices allowing active participation, intense interaction and deep immersion of people in the cultural universe. Immersive technologies, such as 3D, augmented reality or virtual reality, or even gamified applications (for example, the Fabre museum's Fabre & the City app), present a major interest in this respect, facilitating the appropriation of a heritage or a work by the public. The development of interactive devices has proved useful for certain specific audiences. For example, tactile devices for children facilitate learning, increase their level of interest and most often initiate a dialog with the works (Debenedetti et al. 2009). For adolescents, interactive devices also allow for the appropriation of information and codes of behavior during a museum visit, notably through a decrease in spatiotemporal constraints, a slower visit and one that is enriched on an emotional level (Petr and N'Gary 2014).

Finally, not all audiences have the same motivation in the field of art and culture. Aiming to reach the entire population is certainly utopian. Certainly, in France, for example, the taste and the appetite for culture are relatively high within the population, with many French people developing autonomy, skills and social relationships within the cultural universe. However, many also make no effort to approach culture, despite the numerous means of access offered to them. Digital technology will not be the ideal solution for changing this reality, as social and educational determinants are probably too strong. However, without a prior engagement with culture, audiences cannot access a higher level of knowledge and culture and extract the full value of the cultural experience.

9.5. Public engagement or disengagement

The expansion of technologies brings into question the production of experience in the cultural domain. In the experiential framework, it is no longer a question of consuming, but of "living an experience" (Bourgeon and Filser 1995). Thus, the audience becomes the unit of production of experience through the processing, interpretation and appropriation of the products and services offered on the market. Vézina (1999) proposes four characteristics that allow us to better understand experiential culture: (i) the public is much more than a public, and it must be studied as a citizen and as a member of a community, a tribe and a family; (ii) the public acts within situational contexts, which may mean that an individual will not necessarily behave consistently.

On the contrary, their behavior will depend on situational, specific, contingent variables; (iii) the public is looking for meaning, and it does not simply process information as the cognitivist approach would suggest. Experience is a source of meaning; (iv) culture is not limited to a purchase, and it is necessary to consider what precedes and follows the purchase. The public seeks to immerse itself in experiences rich in emotions and far removed from their daily lives through re-enchantment (Badot et al. 2015). The experience is therefore the result of co-creation (Filser 2002) between the museum, its territory and its residents (see Box 9.2).

Digital devices at the Fabre Museum (continued)

Since 2018, the Fabre Museum has been deploying its digital strategy outside its walls with the Fabre & the City application: two seasons of an app offering trails for discovering Montpellier, linked to the history of the Fabre Museum. These apps allow the museum to act as part of its territory and combine with the urban digital experiments carried out by the smart city (Montpellier Metropolis service). These geolocated trails combine new forms of narration (comics, video sequences, scripted trails) and gamification (escape games in the city for different types of audiences) to update the image of the museum. These applications were the object of various partnerships in the stages of their creation and promotion (tourist office, faculty of medicine), and to ensure their evaluation (Montpellier Management).

Accompanying visitors in discovering the collections

Since 2016, several apps have been developed, coexisting with the original audio guides. These apps have emerged based on needs and opportunities (partnerships with animation schools, skill share programs). They offer thematic trails aimed at target audiences (*Lovebot*, a romantic trail in the form of a chatbot; *Voyage en Méditerranée antique,* an application for families focusing on antiquity; *Fabre caché*, an application allowing you to discover the protected treasures of the graphic art cabinet, etc.). Since gaming is a good way to engage visitors, these devices broadly integrate new gamification trends by offering gamified routes.

Some of these schemes have been qualitatively assessed among samples of the public. These different experiments now make it possible to construct the stages of a rationalized and modernized deployment of digital technology all along the trail through the collections.

Perspectives: when technology hides itself to enhance the encounter with the works

Our mediation strategy is in no way dictated by technological innovation, and we ensure that the devices prepare for (or accompany) the encounter with the works. Today, this trend is being affirmed and technology is hiding itself to enhance the works in construction of the future project of accompaniment throughout the collections. Indeed, by 2025, our audiences will be able to enjoy thematic audio trails thanks to the latest immersive sound technologies and new narratives. Thus, the choice of an exclusively sonic experience, totally fluid, removes the barrier of the screen between the visitor and their discovery of the works. This choice also responds to reflections that have turned toward perspectives of disconnection, where the visit would be felt as an escape, disconnected from usual life as embodied in the smartphone.

Box 9.2. *Embedding the museum in its territory and discovering the city and the museum differently (Anne Le Cabec, head of digital projects, Fabre museum)*

Cultural experience is the product of the interaction between an audience, a work and a situation. The sources of value creation lie in the development and implementation of this interaction. The public is not a passive spectator, but, on the contrary, takes an active part, through its actions and commitment, in the extraction of value, that is, the benefits of its own experience (knowledge, pleasure, entertainment, social links, etc.), and in the co-creation of value with other stakeholders (cultural organizations, artists, etc.).

First, audiences are active in extracting value and investing resources in the experience. According to Le Gall-Ely et al. (2008), "cultural practice is never completely cost-free". Even in a context of zero financial charge, it assumes that individuals invest time and effort in accessing works and engage with them. In museums, the intellectual effort, the time required and the organization of the visit are more significant obstacles than the price. However, without intellectual effort and an investment of time, audience involvement is very limited and does not permit delivery of a valuable experience.

Second, audiences are potentially active in the co-creation of value and may contribute in different ways to the production and diffusion of culture (ideation, production of content and sharing). Through their communication, they reinforce the attractiveness of cultural destinations. Communication is no longer left solely to companies or cultural institutions (Cuomo et al. 2021). Trusting their peers, audiences come together in communities to exchange and share their knowledge and experiences (Buhalis and Foerste 2015). In line with the work of Pine and Gilmore (1999) on the economy of experience, economic and institutional actors can understand the experience as a new positioning tool (Mencarelli 2005) and an opportunity to differentiate themselves (Maffesoli 2007). This logic of co-creation is found at different levels, such as within online communities which constitute an important source of data (Big Data) and offer potential new value for decision makers (Khan et al. 2014).

9.6. Conclusion: smart culture, smart people

If development of a smart culture is at the heart of the social, societal and cultural project of the smart city, does it contribute to its democratization and the emergence of smart people?

Smart culture can only achieve its educational objective if it fulfills, on the one hand, its role as mediator by allowing the greatest number of people to access cultural works and objects, and, on the other hand, the educational transformation of the individual into an enlightened, educated or even emancipated being through the cultural knowledge they acquire.

In the connected world, and even more so in the smart city, the individual is immersed in an informational universe which demands – or even over-demands – their attention and cognitive resources (N'Goala 2016). In this digital economy, where only the individual's time is scarce, platforms employ numerous efforts to alert, demand and "grab" the individual's attention. Far from traditional methods of seduction, digital technology seeks to continually notify and keep the user's attention on this content, in the hope of a presence in their mind. These changes lead to major behavioral transformations, and presently, an individual's average attention span does not exceed eight seconds. Patino (2019) speaks, in this regard, of the "goldfish civilization" to describe the confinement of individuals behind their screens, subject to incessant flows of information and the resulting lack of discernment. Psychologists also point out the existence of new disorders such as the fear of missing out on important information. An individual becoming habituated to these multiple solicitations is led to pursue a frenetic exploration and to continue consultation to satisfy a perpetual need for stimulation. Faced with these zapping behaviors, the development of a smart people is very relative, because these eight seconds of exhibition make it difficult to understand, decipher, analyze and appreciate a work. Digital technology in art and culture may also be a factor of social isolation rather than of sharing. Lehn and Heath (2016) notably highlight the effects of psychological and physical disconnection from the outside world when using digital technologies.

The existence of digital influencers (or opinion leaders) also raises the formidable question of the manipulation of the individual and of a partial, selective and biased culture. Katz and Lazarsfeld's two-step communication theory (1955) has long emphasized the role of influential third parties who mediate the dissemination of information to their communities. This mediation is subjective and oriented and accepted by individuals based on the credibility of the influencer. However, in the era of "post-truth", where false information is truer than life, and of artificial generative intelligence, the individual is more exposed than ever to attempts at manipulation. For example, documentaries about extraterrestrials constructing the pyramids of Giza are numerous, and deep fakes, that is, falsification of images or videos by artificial intelligence, again allow the generation of false information on a large scale, creating a totally false "parallel reality". This influence and "disinfluence" marketing is therefore autonomous and raises fears of cultural manipulation of the individual, from which it will be difficult to escape without the intervention of credible mediator–influencers such as local digital facilitators or recognized speaker–guides.

Beyond that, the link between smart culture and the development of a smart people might be challenged by individuals themselves, if they reduce smart culture to only the hedonic and personal enhancement benefits of digital social networks. Social networks provide many benefits, including that of self-esteem, driven by a form of digital egotism (Hennig-Thurau et al. 2004). The cult of self then takes on its full meaning and smart culture becomes a way to feed it, or even oppose it to someone's friends as a totem of social distinction. This instrumentalization is sometimes performed with the complicity of cultural places, which develop applications and encourage the publication of content on social networks to create visibility. Many visitors post photos, videos and stories with their friends on networks with large audiences (Instagram, TikTok and Facebook), and this user-generated content acts as an implicit standard for the popularity of cultural locations. However, it is not without collateral impact, if sharing on social networks becomes the main motivation of *smart culture*. This is illustrated by the selfization of society, where the art of self-portraiture promotes the staging of the self, where culture is nothing more than a backdrop. At the Louvre, the Mona Lisa is now a prisoner of the many selfie sticks parading in front of her, while the Van Gogh Museum in Amsterdam has banned them altogether, stressing that selfies "prevent the appreciation of art". The creation of smart culture, which connects every place, object and cultural experience to the Internet, might accelerate the instrumentalization of culture as personal valorization, neglecting the educational and transformational purpose of smart culture into smart people.

At the sociological level, the project of a smart people is only an illusion if it only turns social determinisms into digital determinisms. The end of entrance barriers is an important step forward for smart culture. Bourdieu (1979) showed that culture constitutes a capital socially determined by class and its social belonging. This cultural capital is owned, controlled and transmitted by the upper social classes. Smart culture reduces these determinisms by making them accessible to the greatest number. However, at the same time, many individuals express difficulties, even reluctance, regarding technology (N'Goala 2016). Digital illiteracy, that is, "insufficient mastery of the digital skills necessary for any person to independently carry out everyday activities" (Maroun 2022), affects 17% of the French population. How can the individual access smart culture without the basic skills? How can the public access culture through new digital media and derive value from them without sufficient mastery of these tools? For example, the perceived educational, recreational and social value of personalization varies greatly depending on individuals' familiarity with the use of smartphones (Ben Nasr et al. 2017). Furthermore, the cost of the equipment may represent a significant obstacle to its democratization. This brings into question access to digital resources and the risks of digital inequalities listed by Beauchamps (2009) and linked to telecommunications infrastructure networks within territories, across the diversity of territories, to devices in the home (a function of household income), to the rapid obsolescence of equipment, but also to inequalities of use (Reddy et al. 2020). A double exclusion from *smart culture*, both social and digital, is likely. These concerns deserve attention from the architects of the smart city, for example, by improving the perceived (Cauchard and N'Goala 2021) and real digital inclusion of individuals.

So, how can we meet the challenge of the digital transition without altering cultural works and the primacy of creation, without destroying the challenge of mediation (knowledge) in favor of fun but hollow technological gadgets, and without producing new exclusions based on technology (competence and autonomy), which would add to the social determinism already latent in the cultural universe? The social and cultural urgency is therefore to experiment with different systems by combining technological advances with the transformation of uses.

9.7. References

Alexandre, O., Alganb, Y., Benhamouc, F. (2022). La culture face aux défis du numérique et de la crise. *Notes du conseil d'analyse économique*, 70, 1–12.

Badot, O., Bree, J., Filser, M. (2015). Entre réenchantement du quotidien et pouvoir du consommateur : regards sur le marketing dans la culture de consommation. *Économies et sociétés*, 23(1), 7–32.

Beauchamps, M. (2009). L'accessibilité numérique : transformer le risque de renforcement des inégalités numériques en opportunité. *Les Cahiers du Numérique*, 5(1), 101–118.

Bec, A., Moyle, B., Timms, K., Schaffer, V., Skavronskaya, L., Little, C. (2019). Management of immersive heritage tourism experiences: A conceptual model. *Tourism Management*, 72, 117–120.

Ben Nasr, I., Hallem, Y., Lagier, J. (2017). Quel est le rôle de l'application mobile dans la valorisation de l'expérience muséale ? *Revue management et avenir*, 2, 87–108.

Bourdieu, P. (1979). Les trois états du capital culturel. *Actes de la Recherche en Sciences Sociales*, 30(1), 3–6.

Bourgeon, D. and Filser, M. (1995). Les apports du modèle de recherches d'expériences à l'analyse du comportement dans le domaine culturel : une exploration conceptuelle et méthodologique. *Recherche et Applications en Marketing*, 10(4), 5–25.

Buhalis, D. and Foerste, M. (2015). SoCoMo marketing for travel and tourism: Empowering co-creation of value. *Journal of Destination Marketing & Management*, 4(3), 151–161.

Carrozzino, M. and Bergamasco, M. (2010). Beyond virtual museums: Experiencing immersive virtual reality in real museums. *Journal of Cultural Heritage*, 11(4), 452–458.

Carù, A. and Cova, B. (2003). Approche empirique de l'immersion dans l'expérience de consommation : les opérations d'appropriation. *Recherche et Applications en Marketing*, 18(2), 47–65.

Carù, A. and Cova, B. (2006). Expériences de consommation et marketing expérientiel. *Recherche Marketing, Nouvelles Tendances*, 162(3), 99–113.

Cauchard, L. and N'Goala, G. (2021). L'inclusion numérique : se rapprocher du public pour des stratégies plus efficaces. *Management des technologies organisationnelles*, 13(2), 171–181.

Combaluzier, E. (2020). La transformation numérique dans la sphère culturelle et patrimoniale. Internship Thesis, Montpellier Management, Université de Montpellier.

Cova, V. and Cova, B. (2001). *Alternatives Marketing*. Dunod, Paris.

Csikszentmihalyi, M. (1997). *Finding Flow*. Perseus Book, New York.

Cuomo, M.T., Tortora, D., Foroudi, P., Giordano, A., Festa, G., Metallo, G. (2021). Digital transformation and tourist experience co-design: Big social data for planning cultural tourism. *Technological Forecasting and Social Change*, 162, 120345.

De Blas, M.D.M., Bourgeon-Renault, D., Jarrier, E. (2015). Can interactive mediation tools bridge the identity gap between the public and the art museum? *International Journal of Arts Management*, 18(1), 52–64.

Debenedetti, S., Caro, F., Krebs, A. (2009). I'd rather play than look at statues: The experiences of children with art works and interactive devices at an art exhibition. *International Journal of Arts Management*, 11(3), 46–58.

Filser, M. (2002). Le marketing de la production d'expérience : statut théorique et implications. *Décisions Marketing*, 28, 13.

Godin, C. (2011). La culture pour chacun : une nouvelle politique culturelle ? *Cités*, 45(1), 164–168.

Graillot, L. (2005). Réalités (ou apparences ?) de l'hyperréalité : une application au cas du tourisme de loisirs. *Recherche et Applications en Marketing*, 20(1), 43–64.

Guintcheva, G. and Lagier, J. (2019). Quel est l'impact du digital dans la valorisation de l'expérience muséale familiale ? *Management & Avenir*, 5(111), 57–78.

Hennig-Thurau, T., Gwinner, K.P., Walsh, G., Gremler, D.D. (2004). Electronic word-of-mouth via consumer-opinion platforms: What motivates consumers to articulate themselves on the Internet? *Journal of Interactive Marketing*, 18(1), 38–52.

Holbrook, M.B. and Hirschman, E.C. (1982). The experiential aspects of consumption: Consumer fantasies, feelings, and fun. *Journal of Consumer Research*, 9(2), 132–140.

Jarrier, E. and Bourgeon-Renault, D. (2020). L'enrichissement de l'expérience de visite muséale par l'utilisation d'outils interactifs de médiation. *Décisions Marketing*, 97, 87–101.

Jarrier, E., Bourgeon-Renault, D., Belvaux, B. (2019). Une mesure des effets de l'utilisation d'un outil numérique sur l'expérience de visite muséale. *Revue management et avenir*, 2, 107–126.

Katz, E., Lazarsfeld, P.F., Roper, E. (2017). *Personal Influence: The Part Played by People in the Flow of Mass Communications*. Free Press, New York.

Khan, N., Yaqoob, I., Hashem, I.A.T., Inayat, Z., Mahmoud Ali, W.K., Alam, M., Gani, A. (2014). Big data: Survey, technologies, opportunities, and challenges. *The Scientific World Journal*, 2014(1), 1–18.

Lachance, J. and Bruna, Y. (2019). De la convivialité dans la ville à l'ère du numérique. *Revue du MAUSS*, 2, 151–164.

Laurent, S. (2022). Les effets de l'accessibilité et de la valorisation de l'expérience sur l'adoption de pratiques de mobilité douce : une étude empirique au sein des grands espaces urbains. Thesis, Gestion et Management. Université de Montpellier.

Le Gall-Ely, M., Urbain, C., Bourgeon-Renault, D., Gombault, A., Petr, C. (2008). La gratuité : un prix ! *Revue française de gestion*, 6(186), 35–51.

Lehn, D. and Heath, C. (2016). Action at the exhibit face: Video and the analysis of social interaction in museums and galleries. *Journal of Marketing Management*, 32(15/16), 1441–1457.

Lemoine, J.F. and Salvadore, M. (2018). L'impact des usages du smartphone sur l'expérience touristique : le cas de la découverte d'une destination. *Revue management et avenir*, 1, 165–189.

Maffesoli, M. (1990). *Au creux des apparences : pour une éthique de l'esthétique*. Le Livre de Poche, Paris.

Maffesoli, M. (2007). *Le réenchantement du monde. Une éthique pour notre temps*. La Table Ronde, Paris.

Maroun, E. (2022). Illectronisme et illettrisme : la question des compétences minimales pour maîtriser les outils numériques. *Informations sociales*, 205(1), 33–37.

Meijer, A. and Bolívar, M.P.R. (2016). La gouvernance des villes intelligentes. Analyse de la littérature sur la gouvernance urbaine intelligente. *Revue internationale des sciences administratives*, 82(2), 417–435.

Mencarelli, R. (2005). Conceptualisation et mesure de la valeur perçue d'un lieu de consommation. Application au domaine du spectacle vivant. *Actes 21ème congrès international de l'Association Française du Marketing*, Nancy.

Mifsud, M., Molines, M., Cases, A.S., N'Goala, G. (2019). It's MY health care program: Enhancing patient adherence through psychological ownership. *Social Science & Medicine*, 232, 307–315.

Müller, J. and Schreiber, A. (2022). Les sorties culturelles des Français après deux années de Covid-19. *Collection Culture études*, 20.

N'Goala, G. (2016). Le marketing dans un monde connecté, un monde de paradoxes. *Décisions Marketing*, 84(4), 5–18.

Pagès, D. (2020). La démocratisation culturelle et les promesses des médiations culturelles numériques : mirage ou tournant ? *Quaderni*, 99(100), 97–112.

Patino, B. (2019). *La civilisation du poisson rouge : petit traité sur le marché de l'attention*. Grasset, Paris.

Petr, C. and N'Gary, E. (2014). L'œuvre vue à travers une tablette mobile : quid des comportements d'appropriation culturelle et des modes de réception artistique. In *Internet : interactions et interface*, Dang Nguyen, G. and Dejean, S. (eds). L'Harmattan, Paris.

Pieters, R.G.M. and Van Raaij, W.F. (1988). Functions and management of affect: Applications to economic behaviour. *Journal of Economic Psychology*, 9(2), 251–282.

Pine, J. and Gilmore, J.H. (1999). *The Experience Economy*. Harvard Business School, Boston.

Reddy, P., Sharma, B., Chaudhary, K. (2020). Digital literacy: A review of literature. *International Journal of Technoethics*, 11(2), 65–94.

Rivière, A. and Bourliataux-Lajoinie, S. (2017). Les effets des m-services touristiques sur la proposition de valeur d'une ville. *Décisions Marketing*, 85, 101–119.

Sandri, É. (2016). Les ajustements des professionnels de la médiation au musée face aux enjeux de la culture numérique. *Études de communication*, 46(1), 71–86.

Uşakli, A., Koç, B., Sönmez, S. (2017). How social are destinations? Examining European DMO social media usage. *Journal of Destination Marketing & Management*, 6(2), 136–149.

Vézina, R. (1999). Pour comprendre et analyser l'expérience du consommateur. *Gestion*, 24(2), 59–65.

10

Consuming

10.1. Introduction: claims of better consumption in the city of tomorrow

The neologism "smart city" designates a space where the city of tomorrow can be projected (Chopplet 2018). The concept of the smart city as such covers multiple and complex realities whose issues cannot be reduced solely to the question of high-tech solutions: "There is no smart city without smart users" (Orsoni 2015).

In their work, Breux and Diaz (2017) summarize the domains and subdomains commonly attributed to the smart city: technology and data, economy, population, transport and mobility, quality of life and environment, tourist destination and governance.

It is worth noting that consumption of goods and services is generally addressed very little in academic and professional conceptualizations of the smart city or in public strategies on the ground, even in the laboratories of the "cities of tomorrow", such as Songdo (South Korea), Masdar City (United Arab Emirates), UN17 Village (Denmark) or Stockholm Royal Seaport (Sweden).

Consumption in the smart city is dealt more so with by local authorities, under the angles of (i) energy (optimization of energy use: intelligent energy management, efficient public lighting, regulation of heating and air conditioning of buildings, etc.); (ii) natural resources (reduction of resource use: green roofs, green walls, intelligent irrigation systems in green spaces, optimization of water consumption, etc.); (iii) data (data production using sensors and mobile applications, to improve population services and energy efficiency). Felli (2015) also states that today smart cities often correspond to a form of development based solely on certain very

---

Chapter written by Fabien DURIF.

For a color version of all figures in this chapter, see www.iste.co.uk/ngoala/smartcity.zip

circumscribed districts and certain sectors of governance, such as energy and transport.

Figure 10.1. *Illustration of The Sustainable City project in Dubai (© VAN)*

The private real estate project The Sustainable City, from Diamond Developers, is a community of more than 500 homes (villas), including a school, an autism rehabilitation center, recreational and sports facilities (equestrian center, stadium), a mosque, a 15,000 m^2 outdoor shopping center (in which all the shops offer sustainable goods), a farm of more than 3,000 m^2 (11 biodomes) and notably a vertical hydroponic farm. There is also an automatic electronic waste management system (container managed by a mobile app) and easy access to sorting for all materials, including textiles.

The streets are cooled by wind towers. All buildings face north, benefiting from as much shade as possible, and are coated with UV-reflective paint.

More than 80% of the development is car free: there are two parking lots shaded by solar panels, as well as additional spaces for communally owned electric buggies for each villa. The maximum distance between the solar parking lots and the villas is 85 m. Cycling is encouraged throughout the community[1].

Box 10.1. *The Sustainable City project in Dubai by the firm Diamond Developers*

1 To learn more, see: https://www.thesustainablecity.ae/.

Nevertheless, questions of "sustainability" and resilience are at the heart of so-called smart city projects (Adolphe 2022). However, few cities concretely encourage "sustainable consumption" of goods and services by promoting models of a circular economy, a sharing economy, ecological shopping, buying from local businesses or buying locally made goods.

We were able to observe this phenomenon during our sabbatical year in 2022–2023 by performing an analysis of the influence of urban intelligence on the sustainable consumption practices of citizens in 17 cities ranked in the IMD Smart City Index 2023: Copenhagen (4th), Singapore (7th), Stockholm (10th), Abu Dhabi (13th), Seoul (16th), Dubai (17th), Sydney (18th), Auckland (22nd), Brisbane (24th), Melbourne (31st), Gothenburg (36th), Madrid (37th), Paris (46th), Toronto (48th), Busan (49th), Montreal (69th) and Barcelona (75th). One private project that caught our attention was Diamond Developers in Dubai, in which sustainable consumption practices are taken into greater consideration in everyday uses.

However, a smart city aspires to be a more sustainable city that indirectly allows its inhabitants to adopt more environmentally friendly behaviors through the use of data and new information and communication technologies (Breux and Diaz 2017; Keita et al. 2022). Given that the proportion of "eco-sceptics" has recently increased from 37% to 44% of the world's population (Kantar Group 2022), the role that cities could play in the environmental and energy transition should not be overlooked.

The question is therefore how to make citizens' consumption practices effective and sustainable over time through the use of urban intelligence. Several questions will arise: which of the existing infrastructures can be described as "smart"? Do current infrastructures meet all the population's needs in terms of sustainable consumption? Does there exist a formal and comprehensive smart city strategy that includes the dimension of sustainable consumption of goods and services? Is the citizen a central actor in the implementation of more sustainable consumption in this so-called smart city? What will it take for sustainable consumption to become the norm in the city of tomorrow?

10.2. The need to move toward more "sustainable" consumption

The most recent report from the IPCC (Intergovernmental Panel on Climate Change), released in March 2023, highlights unsustainable trends in our modes of production, consumption and lifestyles. Indeed, households' carbon footprint, which corresponds to the greenhouse gas (GHG) emissions generated to meet consumption needs, increases proportionally to increases in current expenditure (Institut de la statistique du Québec 2022).

In France, for example, the carbon footprint of final demand relative to the population is estimated at 9.2 tonnes of CO_2 equivalent in 2022 (Ministère de la transition écologique et de la cohésion des territoires 2022): 30% for household transport, 23% for housing; 22% for food; 10% for purchase of capital goods; and 16% for market and non-market services.

Thus, as the General Commission for Sustainable Development (2023) notes, characteristics and modes of consumption have changed significantly and accentuated their impacts on the environment, requiring changes for more rational consumption. Moreover, in recent years, according to the Fondation pour la nature et l'homme (2023), 77% of French people (81% of women versus 73% of men) say they have changed their lifestyle habits to help preserve biodiversity and the climate.

10.2.1. *A diversity of practices serving more "sustainable" consumption*

Although no standard definition has yet been accepted in the academic literature, it generally emerges that sustainable consumption must ensure the satisfaction of basic needs and a better quality of life for present and future generations (Haider et al. 2022).

Sustainable consumption refers to ways of buying, using and disposing of goods and services that have ecological and social attributes. It is based on a model of rational consumption as well as efficient use of goods and services (Quoquab et al. 2019).

As Barber (2007) points out, sustainable consumption is concerned with raising awareness and changing consumption, values and motivations. It involves the adoption by consumers of different practices which the Responsible Consumption Observatory (OCR) has measured each year in Quebec since 2010 in its responsible consumption barometer:

– Non-purchase: resistance/anti-consumption/boycott, sobriety, voluntary simplicity, de-consumption.

– Purchase: ecological, local, organic, fair trade, social consumption (respect for social manufacturing standards, employee rights, community, etc.), second-hand, respecting animal welfare, etc.

– Post-purchase: recycling, composting, reuse, donation, sharing, rental, etc.

The problem is that making the right choices is becoming an increasingly heavy burden for consumers: 68% of them say that it is "tiring to have to look for what is responsible and what is not to make the right choices" (GreenFlex/ADEME 2023).

10.2.2. *Building pro-social and pro-environmental attitudes to achieve more "sustainable" consumption*

Although today 71% of French people say they feel climate change in their daily lives, and this is even more the case for those under 35 (77%) (Fondation pour la nature et l'homme 2023), the gap between intentions and practices, the so-called green gap, appears to be significant and may be explained by several factors, such as negative stereotypes, financial costs or reduced comfort and practicality (El Haffar et al. 2020; Dekhili et al. 2023). Notably, 52% of French people feel that they do not have the financial means to take action at their level (Fondation pour la nature et l'homme 2023).

As noted by Haider et al. (2022), to achieve more "sustainable" consumption, we must strive to build pro-social and pro-environmental attitudes, leading to behaviors that allow efficient and responsible use of resources and minimal waste. The academic literature has identified various determinants that influence decision-making regarding the purchase of sustainable goods and services and provide possible explanations for certain inconsistencies. For example, the work of Joshi and Rahman (2015), using a systematic review, identifies two broad categories of determinants: (i) individual factors: emotions, habits, perceived behavioral effectiveness, perceived control, personal values and norms, trust in environmental performance and environmental knowledge; (ii) situational factors: price, product availability, social norms and reference groups, product attributes and quality, "sustainable" brand attributes, brand image and certification.

Despite advances in knowledge, it is still difficult to understand why some citizens tend to engage more than others in sustainable consumption practices. In surveys of the general public, individual health, the future of generations to come and the good of the planet often emerge as the main drivers of life changes. But turning the drivers into action is not easy. On this subject, the Center for Behavior & The Environment (BE.Center) of the RARE organization proposes an interesting framework of intervention for cities based around six behavioral levers.

– Information about the desired behavior: it provides step-by-step instructions (training, ways to adopt the behavior) and increase awareness and understanding (meetings to explain the desired behavior and its importance, communicate clearly/concretely/unambiguously and provide feedback).

– Adopt rules that promote or restrict behavior.

– Increase or decrease the perceived cost, time and effort to adopt a behavior: make the desired behavior more practical and accessible (removing obstacles, proposing alternatives) and undesirable behavior more difficult (creating barriers), and give rewards to encourage or reinforce the desired behavior and penalties for undesirable behavior.

– Use emotional messages to stimulate behavior: know how to use emotions that drive actions toward the desired behavior, personalize the message (give a human face to communication campaigns and adapt messages to sections of citizens to make them relevant).

– Leverage the behaviors, beliefs and expectations of others: these make participation in the desired behavior observable (publicize who has engaged in the behavior and provide a way for people to show that they have engaged in the desired behavior), make the desired behavior the perceived norm (promote success stories, spark conversation about shared beliefs and expectations, and highlight the possibility of social sanctions for undesirable behavior) and eliminate excuses for not engaging in the behavior (encourage public engagement and provide visible indicators of desired behavior).

– Change the context in which choices are made: it provides direct attention (make the desired behavior the default option, simplify messages and decisions, streamline complex decisions, provide shortcuts for multi-step behavior) and uses timely messages (target transition moments and changes in habits, provide reminders about the desired behavior and help develop a plan to achieve the desired behavior).

Thus, imagining roadmaps for promoting public services that are likely to improve sustainable consumption patterns and citizens' quality of life, based on deeper human connections, to create a society of conscious individuals could be one of the central roles of the cities of tomorrow.

10.3. Thinking of the city of tomorrow as a living laboratory for sustainable consumption

Cities, which now house the majority of the world's population, are centers of consumption that generate environmental impacts extending far beyond their geographical limits.

Beyond the "smart" infrastructures that may be made available, it is above all the consumption habits and lifestyles of citizens that determine the carbon footprint of cities (Schröder et al. 2019).

Cities can therefore help reduce their environmental footprint by using their territory as a place for experimentation with sustainable consumption solutions for the benefit of their residents.

They thus become real living open-air laboratories for a new and more sustainable form of consumption (Cohen and Munoz 2016).

Figure 10.2. *Example of construction (source: https://www.instagram.com/vanresearchmedia/)*

The *Brisbane Carbon Challenge* is an initiative implemented by the Brisbane City Council (Australia) in collaboration with the Brisbane Sustainability Agency. It consists of calculating the annual carbon footprint of a household using a tool made available online, based on domestic energy consumption (bills), transport habits and household waste eliminated. Based on the score obtained, a list of measures is offered to the household to reduce its carbon footprint and its expenses. As actions are implemented, households can see their progress online[2].

To learn more, see: https://www.youtube.com/watch?v=gKwAxX0B4Ak.

Box 10.2. *Brisbane Carbon Challenge: an example of a lever for sustainable behavioral change*

2 Profiles of the "champion" households are notably promoted on: https://www.sustainablebrisbane.com.au/programs/brisbane-carbon-challenge/champion-households/.

10.3.1. *The city: a "legitimate" actor promoting more sustainable consumption*

Citizens' expectations of policy makers on environmental issues are high. French people want, above all, to appear more sincere (34%), but also to be able to anticipate better, that is to say, plan for the long term (24%) (Fondation pour la nature et l'homme 2023).

Smart cities can both help citizens to develop sustainable behaviors and planning, but also to change some environmental and social behaviors (Khansari et al. 2014) through several specific actions such as:

– Subsidizing products/activities that contribute less to climate change: (i) subsidy programs for buying used, new or electric bicycles are offered in several municipalities (Montreal, Granby, Cowansville and Laval in Quebec; Angers, Nantes, Poitiers, Vannes in France); (ii) provision of free public transportation (Mont-Tremblant in Quebec, Montpellier in France); (iii) financial assistance program for buying and installing a charging station for residential use (Magog in Quebec); (iv) "shop local" program using vouchers or a local currency supported by municipalities (Montreal); (v) program for free access to community gardens to grow vegetables (Montreal); (vi) grant programs for ecological household equipment (Montreal: LED bulbs, wall insulation, electric thermostats, etc.); (vii) grant programs for sustainable hygiene products (Sherbrooke, Trois-Rivières, Repentigny in Quebec: washable diapers, washable sanitary napkins, etc.).

– Prohibiting products/activities that contribute most to climate change: (i) several cities in Europe have set up low-emission zones (LEZs) to restrict access to the least polluting vehicles (Paris, Lyon, Strasbourg, Grenoble, Brussels, Antwerp, Amsterdam, Rotterdam, Utrecht, Milan, Rome, Turin, Naples, Florence, Bari, Madrid, Barcelona, Berlin, Hamburg, Stuttgart, Lisbon, London, etc.); (ii) the use by shops, grocery stores and restaurants of certain single-use plastic items (shopping bags, plates, containers, trays and utensils) has been banned in several cities (Montreal, Paris, San Francisco, Bombay, Shanghai, Beijing, London, etc.); (iii) some cities even ban the sale of plastic water bottles (Concord in the United States, Bundanoon in Australia).

– Increasing or reducing taxation of products/activities based on their positive or negative contribution to climate change: (i) several cities have implemented special environmental taxes to finance their programs (Saint-Hyacinthe in Quebec: levy of $0.0069 for each $100 of property value); (ii) taxes on oil-heated appliances (Laval in Quebec).

10.3.2. *Making the city of tomorrow a space for sharing*

Beyond certain targeted actions, the emergence and diversity of sharing activity in cities could further contribute to a transformation of urban economies toward more sustainable consumption and production. The Sharing Cities-Sustainable Consumption Production Typology of Cohen and Munoz (2016) presents the different categories of sharing in which cities can be involved, in collaboration with associative or private actors: energy (cooperative, purchasing group, etc.), food (community gardens, food sharing apps, shared fridges, shared kitchens, etc.), goods (exchange platforms, local redistribution of objects, repair café, fablab, etc.), mobility (car sharing, bike sharing, carpooling, etc.) and space sharing (coworking, warehouse, parking, etc.).

For Picon (2016), our cities are already "smart" and can become more collaborative through providing fertile ground for inter-individual exchanges. One interesting example we saw was the Genbrug Centers in Copenhagen, Denmark. The city of Copenhagen has already set up a network of 13 Genbrug ("recycling" in Danish) Centers. These places allow citizens to donate items they do not want and others to take them for free.

10.4. Conclusion: challenges linked to sustainable consumption at the territorial level

As Boulanger and Jeannin (2016) point out, the success of the city of tomorrow will not be measured solely by its provision of so-called "smart" services, but more broadly by its functionalist design. Effectively integrating sustainability into the heart of the consumption practices (particularly of goods and services) of a city's residents and accelerating the societal transition will be one of the major challenges. This will notably require the participation of all stakeholders (public, associative and private) and a process of trial and error!

10.5. References

Adolphe, L. (2022). *Integrated Urban Environment Management and Resilience: From the Contemporary City to Sustainable Urbanity*. ISTE Ltd, London, and John Wiley & Sons, New York.

Barber, J. (2007). Mapping the movement to achieve sustainable production and consumption in North America. *Journal of Cleaner Production*, 15(6), 499–512.

Boulanger, N. and Jeannin, H. (2016). Une ville intelligente et humaine. *Annales des mines – responsabilité et environnement*, 84(4), 22–26.

Breux, S. and Diaz, J. (2017). La ville intelligente : origine, définitions, forces et limites d'une expression polysémique. Report, Centre Urbanisation Culture Société, INRS, Montreal.

Chopplet, M. (2018). Smart city : quelle intelligence pour quelle action ? Les concepts de John Dewy, scalpels de ville intelligente. *Quadernir*, 96(2), 71–86.

Cohen, B. and Munoz, P. (2016). Sharing cities and sustainable consumption and production: Towards an integrated framework. *Journal of Cleaner Production*, 134(Part A), 87–97.

Dekhili, S., Durif, F., Merle, A. (2023). Marketing durable : accélérons les transformations ! *Recherche et applications en marketing*, 38(3), 3–6. doi: 10.1177/07673701231177803.

El Haffar, G., Durif, F., Dubé, L. (2020). Towards closing the attitude-intention-behavior gap in green consumption: A narrative review of the literature and an overview of future research directions. *Journal of Cleaner Production*, 275(1), 122556.

Felli, R. (2015). La durabilité ou l'escamotage du développement durable. *Raisons politiques*, 4(60), 149–160.

Fondation pour la nature et l'homme (2023). Notre baromètre inéditn sur l'humeur écologique des Français. Press release, Odoxa [Online]. Available at: https://www.fnh.org/wp-content/uploads/2023/06/CP-SondageOdoxa.pdf.

General Commission for Sustainable Development (2023). Le mode de vie des ménages. République française, notre-environnement [Online]. Available at: https://www.notre-environnement.gouv.fr/themes/societe/article/le-mode-de-vie-des-menages-6065.

GreenFlex/ADEME (2023). Baromètre GreenFlex/ADEME de la consommation responsable : la santé globale, moteur de la consommation responsable ? *GreenFlex* [Online]. Available at: https://presse.ademe.fr/2023/06/16eme-barometre-greenflex-ademe-de-la-consommation-responsable-2023.html.

Haider, M., Shannon, R., Moschis, G.P. (2022). Sustainable consumption research and the role of marketing: A review of the literature (1976–2021). *Sustainability*, 14(7), 3999.

IMD (2023). IMD Smart city Index (SCI): Report 2023. Report, IMD World Competitiveness Center, Lausanne [Online]. Available at: https://imd.cld.bz/IMD-Smart-City-Index-Report-20231.

Institut de la statistique du Québec (2022). Empreinte carbone des ménages au Québec. Document, Gouvernement du Québec, Quebec [Online]. Available at: https://statistique.quebec.ca/fr/document/empreinte-carbone-menages-quebec/publication/empreinte-carbone-menages-quebec-2018.

Kantar Group (2022). Who cares? Who does? Is inflation stalling the green momentum in FMCG? Report, Kantar Group, London [Online]. Available at: https://kantar.turtl.co/story/whocares-who-does-2022-p.

Keita, M.L., Royoux, D., Yengué, J.L. (2022). Les enjeux de la ville intelligente et du territoire : une articulation à construire entre les savoir-faire techniques et l'ingénierie sociale. *Droit et ville*, 93(1), 219–235.

Khansari, N., Mostashari, A., Mansouri, M. (2014). Impacting sustainable behavior and planning in smart city. *International Journal of Sustainable Land Use and Urban Planning*, 1(2), 46–61.

Ministère de la transition écologique et de la cohésion des territoires (2022). La décomposition de l'empreinte carbone de la demande finale de la France par postes de consommation : transport, alimentation, habitat, équipements et services. Working Document, Ministère de la transition écologique et de la cohésion des territoires, Paris [Online]. Available at: https://www.statistiques.developpement-durable.gouv.fr/la-decomposition-de-lempreinte-carbone-de-la-demande-finale-de-la-france-par-postes-de-consommation.

Orsoni, F. (2015). La ville intelligente, de l'utopie aux réalités. *Droit et ville*, 2(80), 107–119.

Picon, A. (2016). L'avènement de la ville intelligente. *Sociétés*, 132(2), 9–24.

Quoquab, F., Mohammad, J., Sukari, N.N. (2019). A multiple-item scale for measuring "sustainable consumption behaviour" construct: Development and psychometric evaluation. *Asia Pacific Journal of Marketing and Logistics*, 31(4), 791–816.

Schröder, P., Vergragt, P., Brown, H.S., Dendler, L., Gorenflo, N., Matus, K., Wennersten, R. (2019). Advancing sustainable consumption and production in cities – A transdisciplinary research and stakeholder engagement framework to address consumption-based emissions and impacts. *Journal of Cleaner Production*, 213, 114–125.

11

Moving Around

11.1. Introduction: integrating marketing into mobility policies

As the urban population continues to grow around the world, issues of daily mobility within cities raise major challenges for public stakeholders. The "automobile city", that is, one promoting the widespread use of cars, has dominated urban policy for decades in Western countries. This has resulted in a collective dependence on this mode of transport, negatively impacting communities on several levels: the environment, climate, health and the economy while reinforcing certain sociospatial inequalities within urban territories. This model has become obsolete. While the Covid-19 health crisis has tended to favor individual modes of transport, motorized or otherwise (bicycle, walking), the fight against climate change and the IPCC report (2022) send a clear signal to accelerate the demotorization of households. In this context, promoters of low-carbon mobility can draw from different areas of public action, including demand management.

Furthermore, it increasingly relies on another transition, this time a technological one: that of the smart city, which holds a privileged place for mobility. Policies of demand management propose to act directly on individual behavior, in advance of their travel, to encourage them to adopt new practices. Among these measures, marketing and new information technologies appear as increasingly important levers for action. In this chapter, our objective is to demonstrate the scope of these measures by focusing especially on the development of MaaS (Mobility as a Service) type apps.

After a summary presentation of the field of demand management in mobility, we highlight the contributions of marketing approaches to this field. These measures

---

Chapter written by Sara LAURENT and Florence PAULHIAC SCHERRER.

are based on a deep understanding of users, their ability to change modes and their motivations.

But while it involves thinking about changes in mobility from the point of view of users, the approach also places emphasis on the opportunities for use offered by certain technological innovations. Thus, examining the case of MaaS-type applications, we focus our attention on nudge and gamification techniques. We analyze how these may influence users' motivations and encourage them to adopt "softer" forms of mobility, while highlighting their limits.

The field of personal mobility is experiencing major innovations. The recent introduction of AI (artificial intelligence) into the guidance of transport modes is gradually transforming what is available, using connected and automated vehicles, for example. But so-called smart mobility is also widely deployed through "service" dynamics, which change users' conditions of travel. Thus, new digital technologies are available to users that improve their access to different services for better travel. In terms of demand management, MaaS-type service approaches seem to offer new potentials for influencing users' mobility choices using tailor-made travel proposals adapted to their program of activities.

Through the use of MaaS, users have access to several mobility service options through a single interface to manage their journey just-in-time, according to their specific needs: departure/arrival time, preferred modes of mobility, price, distance, chosen stages of the journey (Hietanen et al. 2014; Jittrapirom et al. 2017). According to Lyons et al. (2019), MaaS solutions may be differentiated in terms of their integration of mobility services at different levels: levels 0–2 require a significant cognitive effort from the user to move from one service (route search) to another (payment), and levels 3–5 offer a more fluid and effortless experience with partial or even total integration of operational services. At these high levels of integration, MaaS solutions facilitate the use of several modes of mobility over the same route, or multimodality. This information can be supplemented by nudge and gamification techniques to easily encourage and reward certain mobility behaviors.

It should be stressed that these (r)evolutions upset the governance of urban transport by promoting public–private partnerships, but also increasing user participation. Furthermore, they are of variable geometry and the experiments in progress raise numerous challenges of feasibility, integration into living environments, but also of ethics and equity.

Box 11.1. *Smart mobility and MaaS*

11.2. Demand management for low-carbon mobility

Issues relating to automobile dependency have featured on the political agendas of cities, and more broadly of regional and national governments, for several decades now. Currently, the collective challenges of the ecological transition and the struggle against climate change highlight the urgent need to support low-carbon mobility. From this perspective, the use of so-called "collective", "soft", or even "active" modes of mobility becomes an issue for reducing the widespread use of individual cars, whether this is chosen by or imposed on populations.

11.2.1. *Acting on mobility: the key role of demand management*

The "automobile city" (Wiel 1999) was born from a congruence between public policies and individual choices. Public stakeholders largely financed the development of freeway infrastructure in cities and between regions, from the 1950s to the 1970s, in Europe as well as in North America. They also implemented economic policies supporting automobile production. Municipal politicians followed suit by promoting the urbanization of the suburbs through the creation of low-density residential areas. These dynamics have been echoed in household aspirations in terms of housing, which has led developers to create a residential supply geared toward access to individual home ownership. These collective and individual dynamics have structured the development and functioning of urban territories. "Centrifugal" urbanization (ever further away from the centers and with low density) has favored the widespread use of the automobile, which has become embedded in the travel habits of urban dwellers. These dynamics have generated a collective "dependence" on the automobile (Dupuy 1999). This mode has appeared the most effective way to access activities scattered across urban space and to link trips together in a day.

For several decades, public policies have relied on a range of measures to reduce this dependence, counter its harmful effects and encourage modal shift. Notably, cities are trying to implement ASI-type strategies (Avoid, Shift, Improve), which may be translated as "avoid" (traveling), "shift" (mode of transport) and "improve" (travel conditions) in order to significantly influence their population's daily mobility practices. To achieve this, several registers of intervention are mobilized in a systematic manner: the development of alternative transport options (for a more diversified choice of modes) in the areas of public transport, but also active transport (such as walking and cycling). Denser and more mixed urbanization, preferably located near sustainable transport networks (to reduce the distances to be covered while offering an alternative to the car).

However, supply-side policies, often with substantial investments, seem insufficient to bring about large-scale and lasting changes in behavior. Urbanization policies also struggle to bring individuals closer to their places of activity and to transform their mobility. In this context, demand management may be used to reorient individuals' mobility (preferred modes of transport, distribution of travel over time and space). The measures aim to act on the determinants of behavior, in advance of travel choices, but also all through their sequence of trips, to modify behavior based on existing transport infrastructures and services.

Thus, demand management is generally perceived as a category of soft action, relatively simple and inexpensive to implement, in contrast to the two other aforementioned categories (transport supply and urbanization), the hard measures, requiring heavy investment and time. Demand management has historically been deployed in various contexts throughout the 20th century (Zoubir 2013). In North America, TDM (Transport Demand Management) measures aim to make existing transport infrastructures (particularly roads, but also public transport) more profitable and improve their efficiency during peak hours to change behavior when traveling between home and work (Black and Schreffer 2010). In Europe, demand management measures are aimed more at changing individual behavior, in the broad sense, toward adopting forms of mobility that are more sustainable and/or more suited to their needs. Marked by their different contexts of origin, these policies converge (Paulhiac Scherrer 2021).

11.2.2. *Incentives or constraints*

Zoubir (2013) emphasizes that demand management measures target the four essential dimensions of individual travel: their reasons for travel (why do they travel?), the times of their travel (when do they travel?), the destinations and distances of the trips (where do they travel?) and the modes of transport used (how do they travel?). By targeting these dimensions, a diverse range of measures becomes available. Some are of an incentive nature, to encourage a voluntary change in mobility, while others, more coercive, attempt to constrain individuals to modify their behavior. Furthermore, these measures are sometimes aimed at a broad audience (the population of a city or territory) or, on the contrary, may target specific groups and thus act on a smaller scale (employees in a given employment area).

There are few comprehensive inventories of these measures (Robinson 1997; Limon 2018; Paulhiac Scherrer 2021). Despite their apparent diversity, however, they can be categorized, distinguishing four groups of measures. Pricing measures propose to change individual choices by means of a change in travel costs or use of modes of transport. An incentive price signal, for example, will offer a preferential rate for taking the bus, while introducing a new urban toll will attempt to discourage the use of cars by imposing a new tax. Measures to modify traffic conditions, for their part, act on the physical and temporal context of travel (sharing of roads according to modes and/or hours), the speeds allowed, the availability of parking, among others, to facilitate or constrain the use of a particular mode. Communication campaigns and the dissemination of information, for their part, aim to enrich knowledge of the alternatives available, their uses, etc., but also to modify beliefs or perceptions relating to certain modes of transport, to modify individual choices. Finally, behavior change support programs are set up in places that generate travel, such as places of employment for example. They offer personalized management of workers' mobility, using several tools, generally incentive-based, and personalized support over a period to facilitate the adoption of new mobility habits.

The scope of these measures is variable. A combination of incentive and coercive measures is generally seen as more promising, as is the definition of target audiences. For example, in terms of modal shift from cars to public transport, the most promising interventions a priori tend to target drivers who are highly motivated to change their behavior, or people forced to adopt new transport habits after moving house ("windows of opportunity" interventions). As we will now see, several types of resistance to these measures may arise among the recipients of said measures.

11.3. Supporting changes initiated by public policies through marketing

Despite the diversity of registers of public policy intervention, it is still difficult to change behavior among users who maintain resistance to some of these interventions. Some users do not change, while others still maintain an unfavorable attitude toward the measures and, by extension, toward active mobility. To support public policies, the integration of a marketing approach appears promising. In fact, it allows us to understand these phenomena of resistance, which limit the scope of interventions. In addition, it allows us to better understand users by using a segmentation strategy.

11.3.1. *Resistance to active mobility policies*

Among the disciplines in the humanities and social sciences that study the psychological mechanisms of users, marketing defines resistance notably in relation to consumer behaviors of rejection and non-purchase of brands. According to Roux (2007), individuals may resist individually, or collectively through boycotts, for example, to oppose perceived pressure in a situation. Through resistance, individuals seek to counteract this pressure by stabilizing their consumption (no change, inertia), or by adopting a negative attitude or behavior regarding this perceived, rather than objective, pressure.

According to the typology of interventions presented above, the more coercive the interventions, restricting users' freedom of choice, the more users may perceive a strong pressure that provokes resistance. Thus, carbon tax policies may foster strong resistance among users who perceive them as ineffective, or even costly on an individual and social level (barriers linked to employment, business competitiveness, etc.) (Carattini et al. 2018; Klenert et al. 2018).

Conversely, more incentive-based demand management policies may be more effective and more easily accepted by users. However, when these policies are justified by the ecological argument, they may be evaluated negatively by those who consider this discourse to be disconnected from their reality, and its practices to be inaccessible; which feeds a negative and caricatured image of so-called responsible users, engaged in ecological practices (de Lanauze and Lallement 2018). Others perceive this discourse to be moralizing, even manipulative, depending on the sender of the message, and involving a risk of greenwashing (Monnot and Reniou 2013). These users may then adopt an unfavorable attitude toward incentives for active mobility, which makes it difficult to adopt these practices. For users with a more favorable attitude, although they perceive benefits from active mobility, they are not always ready to change their practices in favor of active mobility on a daily basis (Laurent 2022).

In other words, the effectiveness of demand management policies may be limited by many physical factors, of course, but also and above all by psychological factors that marketing understands through the lens of user segmentation. According to De Toledo et al. (2022), user segmentation is a serious perspective for studies and tests of interventions for behavior change in mobility, particularly based on user preferences and the longer term stages of change.

11.3.2. *The value of a marketing approach for demand management*

Through a fine segmentation of users, marketing forms homogeneous groups, called "segments", which present common features, based on different criteria: socio-demographic, behavioral and psychological. Thus, the different segments identified can each be the object of an adapted and coherent strategy. In demand management policies, Elgaaied-Gambier and Bertrandias (2022) defend the value of marketing to better understand user psychology, and then to perform a segmentation which accounts for the heterogeneity of users. Such an approach allows for better targeting of interventions, efforts, and thus expenditure to deploy these interventions.

> When we deal with urban mobility, we deal with particularly anchored and routine practices, performed automatically by users in their daily lives. To better understand the benefits that are highlighted to encourage the adoption of active mobility, it is therefore appropriate to segment users based on their habits. Based on a survey carried out in June 2021, among a panel of 883 French users living in urban areas with more than 100,000 residents, we distinguish two main categories of users using a cluster analysis: regular car users (at least 5 days a week) and regular users of public transport. Within these two categories of users, we also distinguish two subcategories: monomodal users, who mainly use a single mode of mobility during the week (i.e. only car or only public transport), and multimodal users, who combine modes during the week (i.e. mainly car plus other modes, or mainly public transport plus other modes).
>
> Among users who mainly take the car (N = 406), single-mode users value the use of public transport for discovering more of their city, for traveling in peace, compared to the car which requires constant concentration. Multi-modal users, already familiar with other modes of mobility, are particularly sensitive to the environmental benefits of active mobility.
>
> Among regular users of public transport (N = 477), single-modal users recognize the benefits of cycling on a daily basis to benefit from a better social image among those around them and to discover more of the urban space. Multimodal users are also sensitive to the enhancement of their social image, but also to the introspection offered by the bicycle, as an individual vehicle compared to public transport. These users, who are already familiar with active mobility, consider cycling to be less efficient than public transport, with the perceived physical effort constituting a barrier to adopting this practice.

Box 11.2. *Encouraging the adoption of soft mobility practices based on types of users (Laurent 2022)*

In cities that have adopted a smart mobility approach, demand management policies may be supported by a large number of digital devices. These may be shared in the urban space (digital kiosk, dynamic display) or personal and carried by users, through their smartphones, to access MaaS mobile applications. MaaS solutions facilitate access to a multitude of mobility options for a multimodal experience aimed at reducing the use of individual cars (Paulhiac Scherrer et al. 2019). This objective is illustrated by the Finnish MaaS solution Whim, which encourages users to compare all available mobility options for reaching their destination (public transport, bicycle, taxis, rental cars, etc.) using a single subscription. In the case of Jelbi, developed by the Berlin Transport Authority, all mobility solutions are integrated, but with a choice architecture arranged to highlight their own services and therefore public transport. Thus, MaaS solutions can promote a personalized and seamless journey with active mobility, relying on engagement mechanisms, well understood by marketing, based on nudge and gamification.

11.4. Effective design of MaaS solutions using nudge and gamification

To encourage active mobility and promote user engagement, mobility service operators and providers can draw inspiration from different mechanisms originating in behavioral theories: nudge theory and gamification. Integrating these mechanisms into MaaS helps create a positive experience to support users toward multimodality on a daily basis. Several experiments have shown their effectiveness and limitations.

11.4.1. *Understanding and distinguishing* nudge *and gamification*

A *nudge* is an intervention or a stimulus designed to influence individuals to make the best decision in their own interests (pro-individual nudges) and/or the general interest (prosocial nudges) (Hagman et al. 2015) with a view toward environmental responsibility (Thaler and Sunstein 2008). A nudge is then a useful behavioral complement to interventions aiming at cognitive change such as education, awareness-raising or persuasion. The nudge allows us to move from intention to action by intervening directly on the choice architecture to change behavior in a predictable manner (Thaler and Sunstein 2008). In nudge theory, developed by Thaler and Sunstein, several mechanisms of engagement can be used. Lehner et al. (2016) distinguish four categories of nudge mechanisms: changes in the physical environment (existing infrastructure, available services), the use of a default choice, the framing of information to highlight the best option in the choice architecture and the use of social comparison.

However, one of the main limitations to the effectiveness of *nudge* lies in its design which guarantees its ethical character: keeping all preexisting options in the choice architecture, even the most harmful ones. Behavior change may thus be compromised to maintain users' freedom of choice. To reinforce the effects of the *nudge*, other mechanisms of engagement from gamification may be relevant. Gamification is the use of game elements in non-game contexts (Deterding et al. 2011). As with *nudge*, gamification aims to encourage behavior change through a positive approach without guilt-tripping or forcing the user. Gamification differs from *nudge* in using rewards such as financial incentives to reinforce behavior. Other mechanisms of engagement through gamification are classified by Koivisto and Hamari (2019): the use of elements of success and progression such as points, challenges or feedback; creation of social bonds through cooperation and/or competition; and immersion in a game world using storytelling and personalization.

Using these two approaches, public policies may rely on several mechanisms to further engage users in active mobility using MaaS, but also the available infrastructure and service offerings (changes in the physical environment).

11.4.2. *State of the art of digital solutions to encourage active mobility*

In the urban and transport sciences literature, no less than 25 experiments were conducted between 2007 and 2019 to test the effects of self-monitoring and feedback, gamification and rewards, social comparison, challenges and goal setting, adaptation to user needs and preferences, and suggestion and cooperation around mobility behavior change.

Experiments mainly took place in Europe (13) (Percues, Tripzoom, SUPERHUB, IPET, Peacox, From5to4, Viaggia Roveretgoto, Streetlife, SaveMyBike, Smart, Optimum, Play&Go, GoEco!). From5to4, SaveMyBike, GreenCommute and Motivate are only accessible through a dedicated website. All of the other experiments offer at least a mobile experience using an app, or even a mobile experience and a website.

What can we learn from the experiments presented in the literature? Tripzoom and QT (Quantified Traveler) confirm the effectiveness of self-monitoring techniques using feedback, with objectives adapted to the needs of users, to change mobility behaviors (Jariyasunant et al. 2015; Poslad et al. 2015). Poslad et al. (2015) also emphasize the importance of user experience with a simple and attractive design. Kazhamiakin et al. (2015), with the Viaggia Roveretgoto application, show the positive impact of gamification on engagement and for promoting active mobility behaviors. In the case of Smart, which also uses gamification and rewards, 44% of the 70 users say they have changed their mobility behaviors and intend to continue. Out of those who did not complete the challenge, 20% said they would have done so with personalized challenges and rewards (Huang et al. 2018).

The SUPERHUB app is one of the most advanced, with a combination of several mechanisms (self-monitoring and feedback, gamification and rewards, challenges and goal setting, and adaptation to user needs). The results are encouraging, with a 14% increase in "active mobility choices" within the space of a month (Gabrielli and Maimone 2013).

However, the results are highly dependent on characteristics related to the user. In the case of MM to encourage users to walk, the results depend on the users' current behavior and their personality traits. Thus, the app encouraged walking more among users who usually walked little (thanks to self-monitoring and feedback) and among users who are competitive (thanks to gamification and rewards). Therefore, different nudge and gamification mechanisms must target defined segments of users based on their habits, needs and preferences.

Mechanisms studied	Digital system created to conduct an experiment (bibliographic references)
Self-monitoring and feedback Tracking performance to help the user achieve their goals	Percues (Reitberger et al. 2007); SUPERHUB (Gabrielli and Maimone 2013); IPET (Meloni et al. 2014); Peacox (Bothos et al. 2014); Moves (Zhao and Baird 2014); QT (Jariyasunant et al. 2015); BikeTogether (Cheng and Lee 2015); EcoTrips (Park et al. 2017); MM (Nakashima et al. 2017); SaveMyBike (Petri et al. 2018); Motivate (Myrovali et al. 2018); Optimum (Anagnostopoulou et al. 2018); Play&Go (Ferron et al. 2019); GoEco! (Cellina et al. 2019)
Gamification and rewards (Virtually) rewarding behaviors to encourage the user to adopt them in the long term	UbiGreen (Froelich et al. 2009); Green Daily Guide (Bliznuyk 2011); SUPERHUB (Gabrielli and Maimone 2013); From 5to4 (Castell et al. 2014); IPET (Meloni et al. 2014); Peacox (Bothos et al. 2014); ViaggiaRoveretogo (Kazhamiakin et al. 2015); E-mission (Shankari et al. 2015); QT (Jariyasunant et al. 2015); Streetlife (Kelpin et al. 2016); MM (Nakashima et al. 2017); Smart (Huang et al. 2018); SaveMyBike (Petri et al. 2018); GreenCommute (Wu et al. 2018); Optimum (Anagnostopoulou et al. 2018); Play&Go (Ferron et al. 2019); GoEco! (Cellina et al. 2019); BetterPoints (Bowden and Hellen 2019)
Social comparison Allowing the user to compare their performance with others	Ecoisland (Takayama et al. 2009); MatkaHupi (Jylhä et al. 2013); From5to4 (Castell et al. 2014); IPET (Meloni et al. 2014); Peacox (Bothos et al. 2014); Moves (Zhao and Baird 2014); Tripzoom (Poslad et al. 2015); QT (Jariyasunant et al. 2015); BikeTogether (Cheng and Lee 2015); EcoTrips (Park et al. 2017); MM (Nakashima et al. 2017); GreenCommute (Wu et al. 2018); Optimum (Anagnostopoulou et al. 2018); GoEco! (Cellina et al. 2019)
Challenges and goal setting Providing challenges and setting goals to encourage the user to progress	Ecoisland (Takayama et al. 2009); Green Daily Guide (Bliznuyk 2011); SUPERHUB (Gabrielli and Maimone 2013); Peacox (Bothos et al. 2014); Tripzoom (Poslad et al. 2015); E-mission (Shankari et al. 2015); Play&Go (Ferron et al. 2019)
Adaptation to user needs Adapting feedback and suggestions according to user needs and preferences, context of use and other factors relevant to the individual	Ecoisland (Takayama et al. 2009); MatkaHupi (Jylhä et al. 2013); SUPERHUB (Gabrielli and Maimone 2013); ViaggiaRoveretogo (Kazhamiakin et al. 2015)
Suggestion Suggesting a relevant option to the user at the right time	MatkaHupi (Jylhä et al. 2013); IPET (Meloni et al. 2014); Optimum (Anagnostopoulou et al. 2018)
Cooperation Motivating the user to adopt a behavior by leveraging the willingness of human beings to cooperate	BikeTogether (Cheng and Lee 2015)

Table 11.1. *Nudge and gamification mechanisms studied in the literature on digital solutions to change mobility behaviors*

11.4.3. *Enhancing the effectiveness of nudge and gamification through fine user segmentation*

To strengthen the effectiveness of interventions based on nudge and gamification, a fine segmentation of users reduces the risk of resistance and, conversely, increases the chances of acceptability of interventions and engagement of users in active mobility.

Indeed, as the *nudge* has the objective of creating individual and/or collective interest in behavior change, its effectiveness is limited by the sensitivity of users to acting for themselves and/or for others. This is notably the case for interventions that use the ecological argument for behavior change. Not all users are concerned about the environment, and even where this concern exists, it is not great enough to motivate them to change their mobility behavior on a daily basis. Users face a social dilemma between what is comfortable and practical for them and what is good and virtuous for the planet. According to Lehner et al. (2016), individuals do not necessarily aspire to reduce their environmental impact. To change their behavior, it would be better to test messages about "doing the right thing" for their health, or "acting appropriately" according to social norms. Lehner et al. (2016) advocate segmenting the population with socio-demographic and behavioral data to personalize the message based on the individual and the context.

Furthermore, the effectiveness of gamification-based interventions depends on the users' interest in being "players". Indeed, several typologies of players have been defined in the literature, particularly in the educational sciences, to know which gaming experience to design based on learner profiles. Segmentation thus makes it possible to choose the most suitable mechanisms based on user preferences, but also to anticipate non engagement from users who are insensitive to elements of play. This reduces the risks, especially the economic risks, of implementing gamification-based interventions with financial incentives.

11.5. Conclusion

A marketing approach can effectively contribute to enriching the domain of public demand management policies. Indeed, by producing detailed knowledge of demand, this approach identifies effective levers to be activated according to different segments of users, for example, in the case of MaaS solutions. However, can integration of a user-centered approach, at a micro scale, contribute to enriching urban collective action in favor of decarbonized mobility, at a macro scale? While individual-centered approaches are increasingly useful for better understanding mobility trade-offs and levers of change, the question concerns the scaling up of the promoted solutions. Thus, MaaS solutions, far from having reached their full maturity, must still demonstrate their relevance in various aspects.

While a marketing approach aims at an individualized approach with a "zoom-in" on users, public policies must, for their part, develop interventions to ensure equity for all in accessing sustainable mobility. One strength of MaaS solutions is that they can be addressed to specific target audiences, but also that they can be deployed at different territorial, local or regional, or even national levels (Andersson et al. 2020). However, experiments have been aimed more at user segments in dense urban areas, which are already in close proximity to public transport and quality car-sharing services (Wright et al. 2020). Furthermore, user access these digitally delivered MaaS solutions is still uneven, even in urban areas. While the development of MaaS appears to be a relevant solution to promote active mobility, digital inclusion of users is becoming a major issue for public policy, at the risk of reinforcing the phenomenon of the digital divide. Not all user segments are equal in their use of digital technology, which marks the limit of a techno-centric approach to managing mobility demand. Finally, *nudge* and gamification techniques, implemented through better user segmentation, are also still in the experimental phase. The long-term effectiveness of these techniques remains under study. Does their effectiveness have an "expiry date"? And when there is a change in behavior, is it sustainable? Nothing is less certain.

11.6. References

Anagnostopoulou, E., Bothos, E., Magoutas, B., Schrammel, J., Mentzas, G. (2018). Persuasive interventions for sustainable travel choices leveraging users' personality and mobility type. In *Persuasive Technology: 13th International Conference*, Waterloo, 229–241. Springer, Cham.

Andersson, A., Winslott Hiselius, L., Berg, J., Forward, S., Arnfalk, P. (2020). Evaluating a mobility service application for business travel: Lessons learnt from a demonstration project. *Sustainability*, 12(3), 783.

Black, C.S. and Schreffler, E.N. (2010). Understanding transport demand management and its role in delivery of sustainable urban transport. *Transportation Research Record*, 2163(1), 81–88.

Bliznyuk, A. (2011). Green Daily Guide. Easier environmentally friendly transportation with the help of mobile technologies. In *2011 International Conference on Collaboration Technologies and Systems* (CTS). IEEE, Philadelphia, 612–617.

Bothos, E., Mentzas, G., Prost, S., Schrammel, J., Röderer, K. (2014). Watch your emissions: Persuasive strategies and choice architecture for sustainable decisions in urban mobility. *PsychNology Journal*, 12(3) [Online]. Available at: https://publications.ait.ac.at/en/publications/watch-your-emissions-persuasive-strategies-and-choice-architectur.

Bowden, H. and Hellen, G. (2019). A data driven, segmentation approach to real world travel behaviour change, using incentives and gamification. In *Towards User-Centric Transport in Europe: Challenges, Solutions and Collaborations*, Müller, B. and Meyer, G. (eds). Springer, Cham.

Carattini, S., Carvalho, M., Fankhauser, S. (2018). Overcoming public resistance to carbon taxes. *Wiley Interdisciplinary Reviews: Climate Change*, 9(5), e531.

Castell, N., Liu, H., Kobernus, M., Berre, A.J., Noll, J., Cagatay, E., Gangdal, R. (2014). Mobile technologies and personalized environmental information for supporting sustainable mobility in Oslo: The Citi-Sense-MOB approach. In *Proceedings of the 28th Conference on Environmental Informatics – Informatics for Environmental Protection, Sustainable Development and Risk Management*. EnviroInfo, Oldenburg.

Cellina, F., Bucher, D., Mangili, F., Veiga Simão, J., Rudel, R., Raubal, M. (2019). A large scale, app-based behaviour change experiment persuading sustainable mobility patterns: Methods, results and lessons learnt. *Sustainability*, 11(9), 2674.

Cheng, Y.M. and Lee, C.L. (2015). Persuasive and engaging design of a Smartphone App for cycle commuting. *mUX: The Journal of Mobile User Experience*, 4(1), 1–5.

De-Toledo, K.P., O'Hern, S., Koppel, S. (2022). Travel behaviour change research: A scientometric review and content analysis. *Travel Behaviour and Society*, 28, 141–154.

Deterding, S., Dixon, D., Khaled, R.., Nacke, L. (2011). From game design elements to gamefulness: Defining "gamification". In *Proceedings of the 15th International Academic MindTrek Conference: Envisioning Future Media Environments*. ACM, New York, 9–15.

Dupuy, G. (1999). *La dépendance automobile : symptômes, analyses, diagnostic, traitements*. Anthropos, Paris.

Elgaaied-Gambier, L. and Bertrandias, L. (2022). Environmental regulations and awareness-raising campaigns: Promoting behavioral change through government interventions. In *Marketing for Sustainable Development: Rethinking Consumption Models*, Dekhili, S. (ed.). ISTE Ltd, London, and John Wiley & Sons, New York.

Ferron, M., Loria, E., Marconi, A., Massa, P. (2019). Play&Go, an urban game promoting behaviour change for sustainable mobility. *Interaction Des. Archit. J.*, 40, 24–25.

Froehlich, J., Dillahunt, T., Klasnja, P., Mankoff, J., Consolvo, S., Harrison, B., Landay, J.A. (2009). UbiGreen: Investigating a mobile tool for tracking and supporting green transportation habits. In *Proceedings of the Sigchi Conference on Human Factors in Computing Systems*. ACM, New York, 1043–1052.

Gabrielli, S. and Maimone, R. (2013). Digital interventions for sustainable urban mobility: A pilot study. In *Proceedings of the 2013 ACM conference on Pervasive and ubiquitous computing adjunct publication*. ACM, New York, 119–122.

Hagman, W., Andersson, D., Västfjäll, D., Tinghög, G. (2015). Public views on policies involving nudges. *Review of Philosophy and Psychology*, 6(3), 439–453.

Hietanen, S. (2014). Mobility as a service. *The New Transport Model*, 12(2), 2–4.

Huang, B., Thomas, T., Groenewolt, B., Fioreze, T., Van Berkum, E.C. (2018). The effect of incentives to promote cycling: a mobility living Lab. In *Transportation Research Board 97th Annual Meeting*, Washington, D.C.

Jariyasunant, J., Abou-Zeid, M., Carrel, A., Ekambaram, V., Gaker, D., Sengupta, R., Walker, J.L. (2015). Quantified traveler: Travel feedback meets the cloud to change behavior. *Journal of Intelligent Transportation Systems*, 19(2), 109–124.

Jittrapirom, P., Caiati, V., Feneri, A.M., Ebrahimigharehbaghi, S., Alonso-González, M.J., Narayan, J. (2017). Mobility as a service: A critical review of definitions, assessments of schemes, and key challenges. *Urban Planning*, 2(2), 13–25.

Jylhä, A., Nurmi, P., Sirén, M., Hemminki, S., Jacucci, G. (2013). Matkahupi: A persuasive mobile application for sustainable mobility. In *Proceedings of the 2013 ACM Conference on Pervasive and Ubiquitous Computing Adjunct Publication*. ACM, New York, 227–230.

Kazhamiakin, R., Marconi, A., Perillo, M., Pistore, M., Valetto, G., Piras, L., Perri, N. (2015). Using gamification to incentivize sustainable urban mobility. In *2015 IEEE First International Smart Cities Conference (ISC2)*, Guadalajara.

Kelpin, R., Giesel, F., Heinrichs, M. (2016). STREETLIFE field trials – Applied gamification approaches as a key to more sustainable mobility behaviour. *In European Transport Conference 2016*. Association for European Transport (AET), Berlin-Adlershof [Online]. Available at: https://elib.dlr.de/105939/.

Klenert, D., Mattauch, L., Combet, E., Edenhofer, O., Hepburn, C., Rafaty, R., Stern, N. (2018). Making carbon pricing work for citizens. *Nature Climate Change*, 8(8), 669–677.

Koivisto, J. and Hamari, J. (2019). The rise of motivational information systems: A review of gamification research. *International Journal of Information Management*, 45, 191–210.

de Lanauze, G.S. and Lallement, J. (2018). Mieux comprendre l'image du consommateur responsable : de la personne idéale aux stéréotypes négatifs. *Décisions Marketing*, 90(2), 15–34.

Laurent, S. (2022). Les effets de l'accessibilité et de la valorisation de l'expérience sur l'adoption de pratiques de mobilité douce : une étude empirique au sein des grands espaces urbains. PhD Thesis, Université de Montpellier, Montpellier.

Lehner, M., Mont, O., Heiskanen, E. (2016). Nudging – A promising tool for sustainable consumption behaviour? *Journal of Cleaner Production*, 134(Part A), 166–177.

Limon, T. (2018). Management de la mobilité, synthèse bibliographique : cadre et définitions, outils, études de cas. Report, Ministère de la Transition écologique et solidaire, Paris [Online]. Available at: https://www.ecologique-solidaire.gouv.fr/sites/default/files/161109_Maitrise%20de%20la%20demande_Rapport%20COP%2022.pdf.

Lyons, G., Hammond, P., Mackay, K. (2019). The importance of user perspective in the evolution of MaaS. *Transportation Research Part A: Policy and Practice*, 121(Part A), 22–36.

Meloni, I., Sanjust, B., Delogu, G., Sottile, E. (2014). Development of a technological platform for implementing VTBC programs. *Transportation Research Procedia*, 3, 129–138.

Meyer, M.D. (1999). Demand management as an element of transportation policy: Using carrots and sticks to influence travel behavior. *Transportation Research Part A: Policy and Practice*, 33(7/8), 575–599.

Monnot, E. and Reniou, F. (2013). "Ras le bol d'entendre parler d'écologie!" : comprendre la contestation des discours écologiques par les consommateurs. *Décisions marketing*, 71(3), 93–109.

Myrovali, G., Tsaples, G., Morfoulaki, M., Aifadopoulou, G., Papathanasiou, J. (2018). An interactive learning environment based on system dynamics methodology for sustainable mobility challenges communication and citizens' engagement. In *International Conference on Decision Support System Technology*. Springer, Cham, 88–99.

Nakashima, R., Sato, T., Maruyama, T. (2017). Gamification approach to smartphone-app-based mobility management. *Transportation Research Procedia*, 25, 2344–2355.

Park, H., Sanguinetti, A., Castillo Cortes, G. (2017). EcoTrips: Leveraging co-benefits and metaphorical metrics in a mobile app to promote walking and biking for short trips. In *Design, User Experience, and Usability: Designing Pleasurable Experiences: 6th International Conference, DUXU 2017*. Springer, Cham.

Paulhiac Scherrer, F. (2021). La gestion de la demande : revue des pratiques. *Cahier In.SITU*, 5 [Online]. Available at: https://chaireinsitu.esg.uqam.ca/nouvelles/cahier-in-situ-n5-la-gestion-de-la-demande-revue-des-pratiques/.

Paulhiac Scherrer, F., Houde, M., Schwach, J. (2019). Étude sur les facteurs et politiques de mobilité durable. *Cahier In.SITU*, 4 [Online]. Available at: https://chaireinsitu.esg.uqam.ca/nouvelles/publicationsrecentes/cahiersinsitu/cahier-in-situ-n4-etude-sur-les-facteurs-et-politiques-de-mobilite-durable/.

Petri, M., Pratelli, A., Farina, A., Buffi, A., Franchina, V., Nepa, P., Iasio, C. (2018). The SaveMyBike Project: ITS technologies and rewarding policies to improve sustainable mobility in cities. In *Proceedings of 7th Transport Research Arena TRA 2018*, Vienna.

Poslad, S., Ma, A., Wang, Z., Mei, H. (2015). Using a smart city IoT to incentivise and target shifts in mobility behavior – Is it a piece of pie? *Sensors*, 15(6), 13069–13096.

Reitberger, W., Ploderer, B., Obermair, C., Tscheligi, M. (2007). The percues framework and its application for sustainable mobility. In *Persuasive Technology: Second International Conference on Persuasive Technology, PERSUASIVE 2007*, Palo Alto. Springer, Berlin, Heidelberg.

Robinson, R. (1997). Transportation demand management in Canada: An overview. *Energy policy*, 25(14–15), 1189–1191.

Roux, D. (2007). La résistance du consommateur : proposition d'un cadre d'analyse. *Recherche et applications en marketing*, 22(4), 59–80.

Shankari, K., Park, J., Gadgil, T., Katz, R.H., Culler, D.E. (2015). Information display for societal problems: Data, game, or choice? Report, University of California, Berkley.

Sunstein, C.R. (2015). *Choosing Not to Choose: Understanding the Value of Choice*. Oxford University Press, Oxford.

Takayama, C., Lehdonvirta, V., Shiraishi, M., Washio, Y., Kimura, H., Nakajima, T. (2009). Ecoisland: A system for persuading users to reduce CO2 emissions. In *2009 Software Technologies for Future Dependable Distributed Systems*. IEEE, Tokyo.

Thaler, R.H. and Sunstein, C.R. (2008). *Nudge: Improving Decisions About Health, Wealth, and Happiness*. Yale University Press, New Haven.

Wiel, M. (1999). *La transition urbaine ou le passage de la ville-pédestre à la ville motorisée*. Editions Mardaga, Paris.

Wright, S., Nelson, J.D., Cottrill, C.D. (2020). MaaS for the suburban market: Incorporating carpooling in the mix. *Transportation Research Part A: Policy and Practice*, 131(2), 206–218.

Wu, S., Bai, Q., Sengvong, S. (2018). GreenCommute: An influence-aware persuasive recommendation approach for public-friendly commute options. *Journal of Systems Science and Systems Engineering*, 27(2), 250–264.

Zhao, J. and Baird, T. (2014). 'Nudging' active travel: A framework for behavioral interventions using mobile technology. In *Transportation Research Board 93rd Annual Meeting*. Transportation Research Board, Washington, D.C.

Zoubir, A. (2013). Management de la Mobilité et pistes pour la pérennisation de la mobilité durable. Quelles voies possibles de régulation ? PhD Thesis, Université Lumière-Lyon II, Lyon.

12

Traveling

12.1. Introduction: technology enhancing the hospitality experience at the heart of a *smart destination*

BENELUS is a city that might be described as exceptional. Imagine disembarking in BENELUS, a city where everything speaks to you and smiles at you! No, you have not arrived in *Alice in Wonderland*, but in a connected smart city. Surprising and authentic at the same time, BENELUS gives you the information you look for, whatever your language. Do you need an address or directions? Do not panic, the city is equipped with interactive screens on every street corner and outside every establishment. Whether looking for an address, reading directions or simply translating a restaurant menu, everything is just a click away. Do you need advice on where to have dinner tonight, or which route to take for your daily jog? Do not panic! Using these same screens, you can access the collaborative platform BienVenuChezVous.benelus, where you can benefit from sincere and spontaneous advice from locals, or simply scan the QR code and download the application. You will thus have the opportunity to interact in real time with nearby residents who can advise you, guide you and even accompany you and share a coffee, a simple walk or your favorite sporting activity with you.

As you can see, in this chapter, we take the reader on a journey to the heart of a *smart destination* to present a new form of tourist experience, tourist experiences improved and enriched through the use of technological tools. These hospitality experiences place encounters and interactions with locals at the heart of tourist experiences which are extraordinary, memorable and sources of well-being (Abbes and Touil 2017). The tourism literature is generating a growing enthusiasm and demand for interactive and participatory experiences, given their strong potential to create value for tourists (Vargo and Lusch 2008; Mathisen 2013; Prebensen et al.

Chapter written by Intissar ABBES and Arnaud MAES.

2012). Tourists are encouraged by the need for socialization and the quest for authenticity to want to integrate into the local culture of the destinations they visit.

In this context, digital technologies act as facilitators of these types of experiences and play the role of intermediaries between tourists and the local population. They can add an additional layer to the hospitality experience and increase its scope. Interacting virtually with the local population would thus allow the postmodern tourist to co-construct their experience and take ownership of it according to their needs. In the digital age, the challenge of *smart destinations* therefore lies in their capacity to rethink their models of competitiveness and tourist attractiveness. Throughout the travel experience before, during and after the stay, the challenge is then to offer connected and innovative technological solutions to, on the one hand, accompany tourists during their stay, and, on the other hand, to enrich their hospitality experience without losing its soul. In fact, technologies may also isolate and cut tourists off from any practical and symbolic interactions with local populations. Does not technology encourage us to keep our distance from local cultures and traditions? What uses of these technologies are likely to strengthen or, on the contrary, impair the perceived hospitality of these destinations? Can technology, in itself, convey an image of hospitality? We will try to answer all of these questions in this chapter.

The chapter will begin with a review of the literature on hospitality and its practice in tourism. The emphasis will be on the hospitality of the local population. Then, we will discuss the contributions of digital technology to territorial tourist hospitality. The central role of the locals will also be emphasized. We will then present the notion of virtual hospitality, highlighting its importance for a destination's territorial hospitality. The challenges of technology will be discussed in this regard. The conclusion of this chapter, emphasizing that a hospitable destination must be hospitable for its inhabitants, will present the case of Foxtrot, the application that improves the daily lives of residents and visitors.

12.2. Hospitality and its meanings

The question of hospitality is dealt with in a great variety of disciplinary fields: theology (Bornet 2010), philosophy (Seffahi 2001), sociology (Gheude and Hullebroeck 2007) or even tourism sciences (Florencio 2012; Touzani 2013). All of these fields, despite their diversity, agree on a concept of hospitality that involves an encounter and interaction between a provider and a beneficiary. The notion of stranger and local or host comes up very often in these approaches to hospitality, despite their diversity. In tourism, hospitality is often associated with experiences that are sources of well-being (Pizam and Ellis 1999; Kandampully and Butler 2001; Gilmore and Pine 2002), with meanings related to gift (Milon 2001), sacrifice

(Gotman 2001) and authenticity (Derrida 2001). Lugosi (2008) speaks of communal and emotional experience.

Touil et al. (2014), for their part, highlight five components of the perceived hospitality of a tourist destination: altruism (the desire to devote yourself to others without expecting anything in return), welcome (the way of receiving a person and behaving toward them at the beginning of the encounter), financial disinterest (an act which does not expect any financial compensation), sincerity (an authentic and spontaneous act toward others) and use of the tourists' language (the fact of communicating with tourists in their language).

Recent research on types of tourism experiences and forms of well-being has classified tourism experiences according to their authenticity and the level of interaction of tourists with locals (Abbes and Touil 2017).

Based on 28 semi-structured interviews with French-speaking tourists of different nationalities (French, Belgian and Canadian), the authors identified three types of lived tourism experiences: commercial, co-created exclusively with professionals; non-commercial, co-created exclusively with the local population; or an intermediate form, involving both professionals and locals. These three types of experiences may relate to simple interactions, or involve more active participation on the part of the tourist. Furthermore, it seems that each type of experience and form of participation involves different forms of well-being.

This is how the sense of well-being evoked by the tourists interviewed seems to cover various forms, which may include positive emotions such as stimulation, calm, pleasure, joy and relaxation, when referring to happiness (Massé et al. 1998; Mogilner et al. 2001), as well as dimensions related to psychological well-being such as self-acceptance, positive relationships with others, autonomy, purpose in life, mastery of the environment and personal growth (Ryff and Keys 1995).

The authors of this research conclude that the key point that allows us to describe an experience as extraordinary is its non-commercial character, which involves authentic social relations and the experiences that have most marked tourists are the non-commercial ones in which they actively participated with the locals. These types of experiences, sources of psychological well-being, prove to be significant and memorable (Gilovich et al. 2015). Indeed, the deep, powerful and enduring character of psychological well-being seems to last beyond the boundaries of time and place. This leaves psychological and emotional traces and markers that create the "memory experience" (Arnould et al. 2002). This induces strongly favorable behavioral intentions toward the destination, translating into a firm intention to revisit and the spread of positive word of mouth.

Box 12.1. *Territorial hospitality, an inducer of well-being and a catalyst for loyalty (Abbes and Touil 2017)*

This approach involves residents, as new ambassadors of their region, in promoting their territory. This type of more participatory, and therefore more authentic, tourism places people at the heart of its approach, far from "staging" and other false pretenses (the phenomenon of folklorization) which are likely to distort the tourist's experience and ultimately impair their well-being. Indeed, the local population must be authentic without falling into hyper-authenticity at the risk of appearing artificial or fake (Dion et al. 2012). The tourism literature is generating a growing enthusiasm and demand for experiences of active interaction and participation with the locals, given their strong potential to create value for tourists (Kreziak and Frochot 2011). This approach is part of the new Transformative Consumer Research (TCR) movement, where the goal is no longer simply to satisfy the customer, but also to delight them and optimize their well-being (Abbes and Touil 2017). Through this conceptualization, hospitality is seen as a multifaceted concept going beyond the sole logic of the quality of tourist service (reception, logistics, services on offer, access to information) to integrate the altruistic and authentic behaviors of local populations.

12.3. The contributions of digital technology to territorial tourist hospitality

Cinotti (2011) speaks of territorial tourist hospitality and includes all of the services supporting the experience of encounter between residents and foreign visitors. Traditionally, these supports took the form of brochures, menus or signage. Nowadays, and since the recent health crises, these supports are being reinvented due to both technological progress and the digital fluency of travelers: connected terminals, interactive screens, intelligent dynamic displays to get informed in real time about the weather, traffic, parking spaces, cultural news and even make suggestions depending on the place. For example, a seaside resort may have digital signs in a central square connected to a system that analyzes the foot traffic in that square. Using smart cameras, we may be offered immersive experiences appropriate to the time and weather, such as people having dinner with their family at a beach restaurant. Beyond these technological performances of the dynamic display type aimed at improving usages, future digital services should complement exchanges between tourists and residents (before, during and after the trip) by promoting physical, but also virtual, flows and links.

Both before and during the trip, through the content created on online platforms (blog posts, content sharing, comments on social networks and websites, online reviews, interactive 360° videos, etc.), a traveler can interact with locals and discover the destination. This is becoming possible thanks to the development of immersive technologies, notably the Metaverse. The Metaverse, a subtle combination of virtual reality (VR) and augmented reality (AR), presents countless advantages for the visiting tourist.

However, the most important feature that makes it superior to all other immersive technologies is that it allows interaction between visitors, but also with residents of the virtual destination through their avatars (Akrout et al. 2022), thus creating a form of virtual hospitality. This can create values for the visitor similar to those a tourist might perceive when interacting with locals during an actual trip (Dwivedi et al. 2022).

In addition to meeting and interacting virtually with the local population and other tourists visiting the Meta, the tourist visitor can interact with tourism professionals and project themselves into the experience, as with the airline Qatar Airways, which is launching QVerse, a new VR experience for visitors to the airline's website. The future traveler will thus be able to virtually navigate within the check-in area, inside the aircraft cabin, and interact with a MetaHuman cabin crew for a more interactive digital hospitality experience. In fact, Pagès (2010) discusses digital hospitality. He presents it as an experience based on local digital services that promote cultural exchange and encounters between visitors and locals for more participatory and inclusive tourism.

Participation can be understood in both directions: the tourist who participates in the life of the local populations or the residents themselves, who participate in tourist activities and who are the real actors. The objective of this participatory tourism is to better respond to the aspirations of a new kind of hyperconnected tourist in search of authenticity.

For example, a creative tourism network is a network that works for the promotion and development of participatory tourism by offering a wide range of creative and artistic activities with locals from regions around the world: photography, cinema, audiovisual, music, live performance, fashion, design, cosmetology, cooking classes, etc. In the same spirit, the Mytriplan platform offers discovery of activities with local guides.

Camille, a tourist who prepared her trip to New York using a trip planner, testifies:

> I organized my trip to New York using a great trip planner. Her wise advice and availability allowed me to prepare my trip efficiently. In addition, the program she prepared for me adhered perfectly to my expectations and desires. MyTriplan really is the right plan for a tailor-made trip and taking advantage of advice from locals to discover the city 'off the beaten track' (Camille, on mytriplan.travel/fr/).

Airbnb supports digital hospitality, with:

– 45.6 million Airbnb users out of 70 million travelers worldwide;

– 2.9 million hosts in all regions of the world.

Airbnb Experiences supports digital hospitality, with:

– 90,000 cabins, 40,000 farms and 3,500 castles, 600 boats, 24,000 tiny homes, 2,800 yurts and 2,600 tree houses, 300 lighthouses and 140 igloos;

– 4.9/5: average rating left by travelers for Airbnb experiences in France;

– five times more reservations in one year in France;

– 500 experiences registered in 2016, more than 30,000 available in 2019;

– 90% of experiences get a five-star rating from users.

The TripAdvisor/Viator platform supports digital hospitality, with:

– 435 million reviews and opinions from travelers around the world;

– 6.8 million properties and accommodations in more than 135,000 destinations, including over 1.1 million hotels, B&Bs, and specialty accommodations;

– 830,000 vacation rental listings;

– 730,000 attractions, including 4.2 million restaurants;

– over 80 million photos of travelers;

– over 280 new contributions published every minute;

– 120 million members to explore worldwide;

– on average, nearly 2,600 new messages are published every day on TripAdvisor forums;

– more than 85% of questions asked on TripAdvisor's English-language forums are answered within 24 hours;

– 6.8 million properties and accommodations in more than 135,000 destinations, including over 1.1 million hotels, B&Bs and specialty accommodations.

Box 12.2. *OTAs (Online Travel Agencies) allied with digital hospitality (2019 figures)*

The same is true of the term "experience" that we find behind brands like Airbnb, TripAdvisor, Decathlon and even start-ups like Sportihome, which appeal to anyone looking for authentic and unique experiences in France, for example, visiting a family wine estate in the Bordeaux region or going rock climbing in the Calanques of Marseille. These private platforms aggregate unstructured data available on the web, while tourist offices feed a TIS (tourist information system), thus making it possible to identify farmers, photographers or even sporting figures who want to show off their region in their own way (Abbes and Touil 2017). We also find the international *Greeters* movement which, using dedicated platforms, promotes encounters between tourists and volunteer residents for a guided tour or a simple walk to discover the unusual places of a city or region. Boosted by cult shows *J'irai dormir chez vous* or *Nus et culottés*, accommodation has also been affected by the wave of participatory tourism, with a proliferation of non-commercial accommodation due to the appearance of numerous platforms, notably the couchsurfing platform that has existed since 2004 in the United States.

After the trip, tourists do not hesitate to relate their experiences and share them on different types of platforms. In fact, in addition to platforms which connect tourists and locals, other types of platforms continue to emerge and develop, such as TripAdvisor, Vacanceo or even IgoUgo. These latter instead make connections between tourists to exchange their travel experiences (good or bad) and advise each other. It seems obvious that in the research, consideration and planning phase of journeys, these supports provide great added value in terms of attractiveness and tourist potential and, ultimately, influence the choice to travel. It is thus through comments on this type of collaborative platform and online reviews that the e-reputation of a destination is formed. A connected city should have an e-reputation!

12.4. Locals: new ambassadors for a *smart destination*

Recent research shows that the most beneficial experiences for a destination and its image are positive experiences with the local population (Abbes and Touil 2017). Conversely, the most damaging experiences for a destination and its image are negative experiences with the local population (Abbes et al. 2019). This result provides lessons for tourism professionals, but especially for local authorities and communities about the importance of involving citizens in promoting their territory. In fact, this means raising awareness among citizens about the importance of positive interactions with visiting tourists through awareness campaigns and programs to combat social problems (racism, discrimination and xenophobia). In this way, the state can encourage a conscious population that respects dignity and accepts others. For example, in 2015, the French Minister of Tourism launched a significant campaign to promote French hospitality. French town halls distributed politeness manuals to traders. They even hired "smile ambassadors" to ensure the spread of good humor around the capital's monuments. In these terms, the public authorities succeeded in engaging the locals in a pedagogy of hospitality. Tourist offices posted brochures and messages of goodwill aimed at tourists throughout the city, on public transport, in shops, etc. These messages were written in French and foreign languages. They also encouraged local associations to participate in sporting and cultural events, while recruiting volunteers to train them and connect them with foreign visitors.

In the same vein, in 2017, the Swedish tourist office integrated into its portal services offered by residents to stimulate contact and encounters between the local population and travelers, such as inviting a tourist to their home, guiding them, renting them a room and showing them their daily life. To facilitate contact and interaction between Swedes and tourists, the Meet the Locals platform was created.

In short, supporting connected tourism for better visibility of authentic experiences, as sources of well-being, seems to be the strategy to adopt to promote a destination.

12.5. Virtual hospitality, a reflection of territorial hospitality

A destination that wishes to be perceived as hospitable should work on its virtual hospitality. The latter has been defined by Bataoui and Giannelloni (2019) as "the capacity of a merchant website to welcome its customers into a friendly and community space, allowing interactions based on a form of reciprocity in the exchange, whether commercial and/or non-commercial". This definition could easily be applied to online tourism platforms whose purpose is to promote a destination (official tourist office websites, Meet the Locals (Abbes and Touil 2017)), a region or even an activity (Spotted by Locals, Like a Local, withlocals, Creative Tourism Network, RendezvousCheznous.com and even the regions, with "BB to our friends" (Béthune Bruay)). Developing virtual hospitality involves working on its four dimensions, namely the spatial dimension, the social dimension, the reciprocity dimension and finally the dimension of play (Bataoui and Giannelloni 2019). The spatial dimension refers to the comfort felt by visitors during their experience of using and navigating the platform. The social dimension refers to interaction and exchanges between tourists and tourism professionals or even among tourist visitors themselves (sharing of experiences). The reciprocity dimension goes beyond the simple notion of exchange and interactions between platform actors to ensure that the visiting tourist appropriates the space, identifies with it, and feels "at home" there. They engage with the platform and its community by creating authentic content. These virtual communities are made up of people sharing common interests without geographical constraints, physical interactions or ethical origins. Its members participate actively and repeatedly in the community and build strong emotional ties (Kozinets 1999). It has been shown in several research works that the level of participation of the members of a virtual community is variable and can go from a simple *like*, to a comment, to the creation of content to feed the community (Raies and Gavard-Perret 2011), to real actions of activism and support for the object of the community (a country, a region and an activity) (Cova et al. 2015;

Abbes and Rahmouni 2021). This level of participation depends on the level of commitment to the community of members who nurture a sense of belonging and cultivate a strong emotional bond (Heehyoung et al. 2008).

Virtual communities promote interactions and social exchanges between their members. They form a space dedicated to the exchange and sharing of experiences. Chen et al. (2013) emphasize the notion of aid between members of a virtual community as being the very essence of these tribes: asking for opinions, sharing experiences, using the experiences of others, etc. All this while offering visitors, members or future members a pleasant and entertaining experience using different means of entertainment: competitions, quizzes and small gifts.

In short, to cultivate virtual hospitality, it seems important for tourism platforms to be collaborative in nature, as these gatherings of consumers (tourists and visitors) can constitute strategic tools for building a brand image and ensuring loyalty and engagement (Abbes and Rahmouni 2021) toward the destination.

This result has also been supported by the work of Bataoui et al. (2022) on the role of virtual hospitality in generating positive effects on attitudes and behavioral intentions. However, would not too much connectivity hold the paradoxical risk of breaking the social bond and swamping the tourist visitor in virtual relationships, where they will have "thousands of hosts and not a single friend"?

12.6. Technology and prejudice: the dark side of virtual hospitality

Conversely, technologies may isolate and cut tourists off from any practical or symbolic interaction with local populations. Indeed, leaving your comfort zone, putting aside your screen to go and meet the real world and interact with locals, exposes the tourist to risks and to multiple sources of irritation and discomfort (Abbes et al. 2019). Why should we talk to strangers, sometimes even in a language we do not speak, when the information we are looking for is just a click away?

The relational variation has often been considered as an inherent cost of using technologies (Mencarelli and Rivière 2014). Despite the undeniable interest in digital, mobile and connected tools, these often entail a certain number of sacrifices of a psychological, sensory, reassurance, spiritual and societal nature (Abbes et al. 2020). They are, at the same time, creators and destroyers of values (Lapassouse-Madrid and Vlad 2016). How many tourists spend their time taking pictures of themselves and making live videos and stories instead of enjoying the present moment and immersing themselves in the experience? Ubiquity certainly provides services, but it may degrade the creation of values. However, living in the present moment and meeting real people still and always remains at the heart of our experiences (N'Goala 2016). Digital technology seems to be a formidable tool for isolation, even for "social embedding" (Helme-Guizon and Magnoni 2016), which may dissuade us from going to meet the local population to ask for directions or find out about an address and taking advantage of these fortuitous moments of exchange, interaction and discovery of others. However, it should be noted that new intelligent, interactive, agile and more collaborative technological devices would allow optimization of tourist flows to destinations subject to overtourism, as the Waze application may do. These technological devices redefine the nature of the relationship between tourists and locals and refer to mastery of the social bond (Abbes et al. 2020).

However, we cannot ignore the upheaval that the world experienced at the end of 2022 with the release of ChatGPT, its democratization and the rapid development of generative AI. Since then, we have witnessed new behaviors from tourists. Instead of a passer-by, they ask ChatGPT to find their way or for advice from the hotel receptionist on where nearby to have dinner tonight. A dialog unfolds between the tourist and the machine, as can be done with the tool *Pi, your personal AI*. The result of this exchange seems highly satisfactory for tourists, who obtain answers to these questions and can ask all the questions they want (timetable, access and attendance), in their own time, without fear of disturbing their interlocutor or abusing their patience and hospitality.

Is it more efficient to talk to a robot? Maybe! But whether it is as enriching as talking to a human, we have reservations! Beyond the informational value, other values of an emotional, social and human nature are expected by the tourist to enrich, authenticate and enhance their hospitality experience.

12.7. Technology supporting a more hospitable destination for its residents

A destination described as hospitable is a destination that is also hospitable for its residents; a destination where it is good to live, capable of making its residents happy and ensuring their well-being, which in turn is felt by visiting tourists. In fact, as Bernard Schlink says so well in his novel *Amours en fuite*, "you have to be happy to make others happy". It is in this sense that technology may play a vital role in improving the daily lives of residents. Take the example of Foxtrot, an app initiated by the city of Orléans. It provides a digital service imagined by residents to transform their daily travels into a pleasant experience of discovering heritage (see Box 12.3).

In 2019, as part of a call for projects from the Orléans metropolitan area on the creation of a new heritage discovery service using *Open Data*, the Montpellier company ENEO responded with a project of co-construction with residents of the heritage that surrounds them. The existing marked trails in the metropolis seem to be little known or difficult to access and not very informative.

The idea is to transform daily travel into opportunities to discover the surrounding heritage. The creation of a tool for personalized heritage walks thus appeared to be an adequate response, expected by residents. Foxtrot offers everyone the opportunity to maximize their contact with nature during their daily travels to make brief stops to reconnect with what they see every day without paying attention to it. Directly on Foxtrot or on the metropolitan app, the user accesses the service via a chatbot. The heart of the service is the custom route generator. These routes are not calculated based on the shortest path (for example, as Google Maps suggests), but on the path that maximizes contact with nature. The routes are generated using Open Data from the metropolis of Orléans, as well as data from Open StreetMap.

In addition to the routes, the tool suggests points of interest to discover along the route, so that each journey becomes a unique experience of reconnecting with our environment.

How Foxtrot works

The operating principle is based on the algorithm that calculates the routes and points of interest based on four user options (theme, mode of transport, departure and arrival points). Weighting the environmental and route data maximizes contact with the chosen theme.

In addition, the algorithm is fed by data sources from the Orléans metropolis Open Data (parks, heritage, cycle paths and city atmosphere, etc.), but also databases of fauna, flora, trees, noise and points of scientific and natural interest. Foxtrot never stops improving.

New profiles and modes of travel have been integrated into the application: "agri-rando" to discover local producers and products, tourism to discover historical and architectural heritage.

In addition, improvements in the quality of data related to nature have made it possible to add links to images and external links of the Wikipedia type.

A new collection and integration of data sets have been carried out (local producers, markets, road diagnostics, rides, parks and gardens, nature trails, public toilets, playgrounds, parking arches, bins and benches).

Evaluation of Foxtrot use

From June 7 to September 8, 2021, 3,600 calls to Foxtrot were made during the summer, nearly 800 of them outside the Orléans metropolis, mainly in Paris and in the Centre-Val de Loire. The impact of the tourist office on incentives to use the application can be demonstrated. There were 32 trips made from and 31 trips to the tourist office. More searches were made on the weekends. Foxtrot ultimately makes it better to live in the city and transforms daily journeys into heritage discovery experiences; this is the challenge of the smart city and its future lifestyles.

Box 12.3. *Foxtrot, the app that improves your daily life*

12.8. Conclusion

At the end of this chapter, we have contributed to painting the picture of a smart city by emphasizing the importance of real and virtual hospitality, both for visitors to a destination and for its residents. Technology plays a key role in this. However, a smart city is not a stack of sophisticated technologies that satisfy the smallest desires of visitors and residents at the click of a button. It also, and above all, concerns smart technologies that facilitate meeting and interacting with locals, useful technologies that create links and do not break them while improving the daily lives of residents and their quality of life. It is in this sense that we can consider the hospitable city to be the other face of the smart city.

This is how local authorities should implement a strategy for promoting their territory, which includes a strong societal dimension and places residents at the heart of the project (innovations with and for citizens, awareness campaigns for residents, information and training activities for locals about smart technologies that improve their city's tourist hospitality experience).

12.9. References

Abbes, I. and Rahmouni, M. (2021). Industrie pharmaceutique et communautés virtuelles des consommateurs : quel design pour ces nouvelles plateformes d'engagement du domaine de la santé ? *Innovations*, 66, 177–202.

Abbes, I. and Touil, S. (2017). Formes du bien-être et expériences de consommation : une application au secteur touristique. *Décision Marketing*, 78, 81–98.

Abbes, I., Bouslema, H., Zemni, R. (2019). The dark side of co-creation during the tourism experience: An exploratory study of the different types of ill-being. *Journal of Travel & Tourism Marketing*, 36(8), 928–940.

Abbes, I., Temessek, A., Hallem, Y., Hikkerova, L. (2020). Le rôle de la technologie dans les expériences de coproduction de service : une lecture sociale de la valeur perçue. *Management International*, 24(5), 106–121.

Akour, I., Al-Maroof, R.S., Alfaisal, R., Salloum, S. (2022). A conceptual framework for determining metaverse adoption in higher institutions of gulf area: An empirical study using hybrid SEM-ANN approach. *Computers and Education: Artificial Intelligence*, 3, 100052.

Arnould, E., Price, L., Zinkhan, G. (2002). River magic: Extraordinary experience and the extended service encounter. *Journal of Consumer Research*, 20, 24–45.

Bataoui, S. and Giannelloni, J.-L. (2019). Rendre le commerce électronique plus humain et moins marchand en considérant les sites internet marchands comme des espaces d'hospitalité. *Management & Avenir*, 7(113), 83–103.

Bataoui, S., Gerard, J., Martin-Lacroux, C. (2022). Quand l'hospitalité du site internet d'une PME l'aide à attirer des candidats. *Décisions Marketing*, 106, 73–89.

Bornet, P. (2004). Judaïsme – entre normes religieuses et impératifs éthiques. In *Le livre de l'hospitalité – Accueil de l'étranger dans l'histoire et les cultures*, Montandon, A. (ed.). Bayard, Paris.

Chen, G.L., Yang, S.C., Tang, S.M. (2013). Sense of virtual community and knowledge contribution in a P3 virtual community: Motivation and experience. *Internet Research*, 23(1), 4–26.

Cinotti, Y. (2011). Hospitalité touristique : conceptualisation et études de l'hospitalité des destinations et des maisons d'hôtes. PhD Thesis, Université de Perpignan Via Domitia, Perpignan.

Cova, B., Pace, S., Skalen, P. (2015). Brand volunteering: Value co-creation with unpaid consumers. *Marketing Theory*, 15(4), 465–485.

Derrida, J. (2001). *Papier Machine*. Galilée, Paris.

Dion, D., Sitz, L., Rémy, E. (2012). Légitimité et authenticité des affiliations ethniques : le cas du régionalisme. *Recherche et Applications en Marketing*, 27(1), 60–77.

Dwivedi, Y.K., Hughes, L., Wang, Y., Alalwan, A.A., Ahn, S.J., Balakrishnan, J., Barta, S., Belk, R., Buhalis, D., Dutot, V. et al. (2022). Meta-verse marketing: How the metaverse will shape the future of consumer research and practice. *Psychologie Marketing*, 40(4), 625–862.

Florencio, A.P. (2012). La place de l'hospitalité dans une prestation touristique sur mesure. PhD Thesis, Nouvelle Sorbonne.

Gheude, M. and Hullebroeck, P. (2007). A l'école de l'hospitalité. Études réalisées pour la Ligue de l'Enseignement et l'Education permanente. Study, Brussels.

Gilmore, J.H. and Pine, B.J. (2002a). Customer experience places: The new offering frontier. *Strategy & Leadership*, 30, 4–11.

Gilmore J.H. and Pine B.J. (2002b). Differentiating hospitality operations via experiences. *Cornell Hotel and Restaurant Administration Quarterly*, 87–96.

Gilovich, T., Kumar, A., Jampol, L. (2015). A wonderful life: Experiential consumption and the pursuit of happiness. *Journal of Consumer Psychology*, 25(1), 152–165.

Gotman, A. (2001). *Le sens de l'hospitalité. Essai sur les fondements sociaux de l'accueil de l'autre*. Presses Universitaires de France, Paris.

Heehyoung, J., Olfman, L., Islang, K., Joon, K., Kyungtae, K. (2008). The influence of online brand community characteristics on community commitment and brand loyalty. *International Journal of Electronic Commerce*, 12(3), 57–80.

Helme-Guizon, A. and Magnoni, F. (2016). Intégrité, bénéfices et identification : trois leviers pour booster l'engagement envers la marque sur les réseaux sociaux. *Décisions Marketing*, 84, 95–114.

Kandampully, J. and Butler, L. (2001). Service guarantees: A strategic mechanism to minimise customers' perceived risk in service organisations. *Managing Service Quality: An International Journal*, 11(2), 112–121.

Kandampully, J., Mok, C., Sparks, B.A. (2001). *Service Quality Management in Hospitality, Tourism, and Leisure*. Haworth Hospitality Press, New York.

Kozinets, R.V. (1999). E-tribalized marketing? The strategic implications of virtual communities of consumption. *European Management Journal*, 17(3), 252–264.

Kreziak, D. and Frochot, I. (2011). Co-construction de l'expérience touristique : les stratégies des touristes en stations de sport d'hiver. *Décisions Marketing*, 64, 23–33.

Lapassouse-Madrid, C. and Vlad, M. (2016). Courses connectées : un cas de destruction ou de création de valeur pour les clients et les distributeurs. *Décisions Marketing*, 84, 43–60.

Lugosi, P. (2008). Hospitality spaces, hospitable moments: Consumer encounters and affective experiences in commercial settings. *Journal of Foodservice*, 19(2), 139–149.

Massé, R., Poulin, C., Dassa, C., Lambert, J., Bélair, S., Battaglini, M.A. (1998). Élaboration et validation d'un outil de mesure du bien-être psychologique : L'ÉMMBEP. *Revue Canadienne de Santé Publique*, 89(5), 352–357.

Mathis, E.F., Hyelin, L.K., Uysal, M., Sirgy, J.M., Prebensen, N.K. (2016). The effect of co-creation experience on outcome variable. *Annals of Tourism Research*, 57, 62–75.

Mathisen, L. (2013). Staging natural environments: A performance perspective. In *Advances in Hospitality and Leisure*, Chen, J.S. (ed.). Emerald Group Publishing Limited, Leeds, 163–183.

Mencarelli, R. and Rivière, A. (2014). La participation du client dans un contexte de self-service technologies. *Revue Française de Gestion*, 4(241), 13–30.

Milon, A. (2001). L'étranger dans la figure de l'hospitalité : rôle et place d'une anthropologie pragmatique. In *Lieux d'hospitalité : hospices, hôpital, hostellerie*, Montandon, A. (ed.). Presses Universitaires Blaise Pascal, Clermont-Ferrand, 41–49.

Mogilner, C., Kamvar, S.D., Aaker, J. (2001). The shifting meaning of happiness. *Social Psychological and Personality Science*, 2(4), 395–402.

N'Goala, G. (2016). Le marketing dans un monde connecté, un monde de paradoxes... *Décisions Marketing*, 84(4), 5–18.

Pagès, D. (2010). Les territoires numériques : au-delà de l'information localisée, l'hospitalité au fil des écrans ? *Quaderni*, 72, 101–112.

Pizam, A. and Ellis, T. (1999). Customer satisfaction and its measurement in hospitality enterprises. *International Journal of Contemporary Hospitality Management*, 11, 326–339.

Prebensen, N.K., Woo, E., Chen, J.S., Uysal, M. (2012). Motivation and involvement as antecedents of the perceived value of the destination experience. *Journal of Travel Research*, 52(2), 253–264.

Raies, K. and Gavard-Perret, M.L. (2011). Intention de fidélité à la marque des participants a une communauté virtuelle de marque : le rôle dual de l'engagement. *Recherche et Applications en Marketing*, 26(3), 23–43.

Ryff, C.D. and Keyes, C.L.M. (1995). The structure of psychological well-being revisited. *Journal of Personality and Social Psychology*, 69(4), 719–727.

Seffahi, M. (2001). *De l'hospitalité, autour de Jacques Derrida*. La Passe du Vent, Genouilleux.

Touil, S., Zine-Danguir, S., N'Goala, G. (2014). L'hospitalité locale vectrice de fidélisation envers une destination touristique ? In *Actes du 30ème Congrès de l'Association Française du Marketing*, Montpellier.

Touzani, L. (2013). Le rôle central du choc culturel dans les expériences d'hospitalité touristique. Doctoral Thesis, Universités de Tunis, Grenoble.

Vargo, S.L. and Lusch, R.F. (2008). Service-dominant logic: Continuing the evolution. *Journal of the Academy of Marketing Science*, 36(1), 1–10.

13

Healing

13.1. Introduction: connected care, digital health in smart cities

So-called "smart cities" are being mobilized for innovation in the health sector. Indeed, the process of global urbanization draws attention to cities that are expected to contain larger populations and provide citizens with services in an efficient and humane manner (Solanas et al. 2014). At the same time, the world's population is aging, the number of people suffering from chronic diseases is increasing, and information and communication technologies are ubiquitous in the health sector. Thus, it becomes important to work more seriously on developing a global vision of a new system allowing the acquisition of health information in a "smart", automatic, and transparent way, to make effective decisions (Rathee et al. 2019).

By exploiting the potential of artificial intelligence (AI), *Big Data*, robotics and the Internet of Things (IoT) (Alam 2021), smart cities offer concrete solutions for proactive management of both health emergencies and long-term strategies. These technologies can be used to optimize the management of cities and have great potential for improving health services and, consequently, residents' health, directly or indirectly.

As international examples we can mention South Korea's *Smart City Data Hub*, an urban data analysis platform used to try to contain the coronavirus epidemic by tracing patients with Covid-19 who may have been in contact (Millard 2020). The *National Steps Challenge*, in Singapore, is a physical activity program aiming to encourage its residents to be more active. Each participant receives a connected

---

Chapter written by Blandine CHAPEL and Sarah GARIDI.

device designed to count the number of steps taken each day and help them pay attention to their diet. Bristol, considered a leading smart city in the field of health, has also developed a living lab based on a global vision of health and collaborative approaches, which allow the integration of citizen participation.

In this chapter, we provide an overview of the main areas of knowledge involved in the construction of this new concept. In addition, we aim to present, problematize, criticize and question these new connected health care systems. These systems may offer attractive applications for managing elderly care, fitness and remote monitoring, or even suggest drug treatments for users and enable better organization of the provision of care in the territories.

Thus, there is no doubt that researchers, professionals and users in the fields of electronics, information and communication technologies and medicine must work in close collaboration to strengthen the global and interdisciplinary vision of this health system (Payton et al. 2011).

To support the construction of smart cities, we discuss the main challenges and opportunities that "connected care" would imply across territories and provide our thoughts for future research.

13.2. Embedding digital health in *smart cities*

13.2.1. *Definition and challenges of digital health in* smart cities

Before going into more detail about the digital dimension, it is still important to recall the definition of the term "health". The World Health Organization (WHO) proposed a comprehensive definition in 1945: "Health is a state of complete physical, mental, and social well-being and not merely the absence of disease or infirmity". In this sense, the use of information and communication technologies (ICT) in the field of health is called, in its most frequent usage, "digital health", "e-health" or even "connected health" (Delmotte 2015).

According to IRDES (French Institute for Research and Documentation in Health Economics), digital health involves two major areas: health information systems or hospital information systems (HIS) and telehealth, which includes telemedicine and m-health (for mobile health). Digital health therefore covers a set of solutions for users, professionals and/or healthcare organizations. Figure 13.1 defines its outlines and recipients.

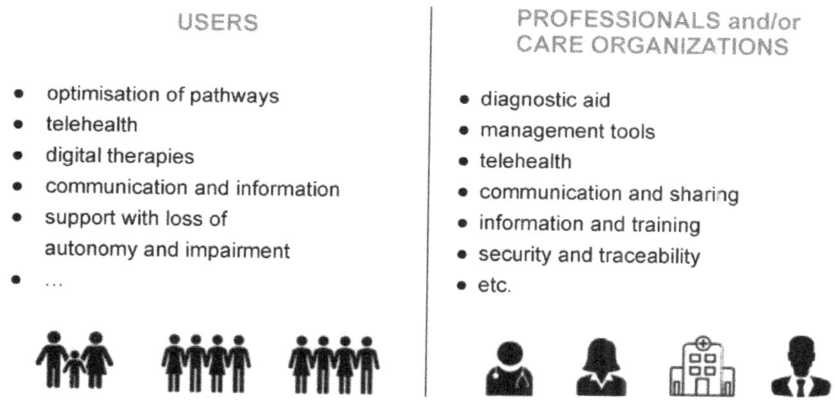

Figure 13.1. *Outlines and recipients of digital health*

In France, innovative projects have begun to appear, linked to health-environment issues, in certain French urban areas such as Rennes. These applications inform city dwellers about pollution levels to encourage them to change their modes of transport.

The influence of digital technology in the health sector is all the more significant as 81% of doctors are convinced of the benefits of digital health for the quality of care: 99 billion euros of potential savings, with mobile solutions in Europe, and three out of four French people prepared to communicate via digital channels (video, SMS, etc.) with their practitioners. Furthermore, as part of the "France 2030" action plan, announced in January 2021, the government is launching a "digital health" acceleration strategy. This acceleration strategy notably aims to encourage the transition from curative, siloed medicine to a more preventative, predictive, personalized approach, and to promote the emergence of innovative solutions for the benefit of the general population and patients in France.

At the same time, cities are becoming increasingly digitalized, accelerated by the concept of smart cities. Political decisions and urban governance are influenced by the emergence of IoT data, which constitute the backbone of smart cities. These data notably give birth to Big Data and the development of analysis using AI models (Allam et al. 2019). New resources then become available to us for understanding the performance of cities, both economically and in terms of health. As a result, we are aware that the deployment of digital technologies is transforming urban life and may have significant effects on physical, mental and social well-being (Mouton et al. 2019).

> Patients are increasingly adopting connected objects, notably to monitor their heart rate, measure their number of steps (and therefore the distances covered), and analyze their sleep. As part of the patient's care pathway, these data may be of great interest to healthcare staff who can also use them to provide remote and continuous monitoring. (Karim Djamai, mobility director, VMware France in "Trends, digital transformation" from the Econom blog, 2017)

Over several years, the scientific community has seen numerous debates around understanding the well-established field of "healthy cities" and the interest in "smart cities". We present some examples here.

The interpretation of sensor data from IoT devices often neglects to include human anatomical and biological data in its various forms, when these could lead to a better understanding of the relationship between people and the city. The combination of these forms of data would allow us to conceive smart cities that are more contextualized, resilient and sustainable, as proposed by Sustainable Development Goal (SDG) 11 and the New Urban Agenda (Allam et al. 2019). However, a study commissioned by the Federation of Private Hospitals for Medicine, Surgery and Obstetrics reveals that 41% of French people have doubts about the security of digital medical data, in line with modest growth in the use of the digital health record "My Health Space". According to the latest figures from February 2023, only 11.5% of users (receivers of state health benefits) have activated their "My Health Space" profile. The issue of data confidentiality and access by third parties comes up regularly, and is also fueled by incidents reported in the media, even though France is protected by several regulations, laws and directives relating to data management, such as the General Data Protection Regulation (GDPR), the existence of the CNIL (French National Commission on Informatics and Liberty) and the mandatory use of health data hosts.

The failure of recent data-sharing initiatives also demonstrates the importance of informed understanding for investment in this area of health (Kalkman et al. 2022). Thus, health research still needs to identify favorable conditions that integrate the values, needs and interests of patients and the public.

Finally, population health and equity are still largely absent from political and academic discussions about "smart cities". Some authors use the concept of "digital divide" or "digital inclusion" (Cauchard 2023) which is, on the one hand, associated with the individual level and the idea of socioeconomic gradients, and, on the other hand, with the level of organizations involved in the promotion of health, the fight against inequalities and responding to the major challenge of access to digital technologies and data (Mouton et al. 2019).

The future of health services is focused on providing people with a comprehensive service, including many factors that affect their health. The implementation of digital technology in this sector seems to foreshadow a real-time analysis, in which doctors, researchers and other interested parties could make more informed decisions, while offering patients greater control over their own medical monitoring (Green and Vogt 2016; Pulciani et al. 2017).

Thus, digital health bears numerous expectations and hopes, whether on the part of the medical world or on the part of patients, for simplifying access to care, overcoming the problems of medical demographics, supporting the shift to outpatient care, promoting the patient's direct participation in their care or supporting innovation through the management of so-called "real-life" data to advance on "4P" precision medicine (Lindenmeyer 2022). These hopes remain nuanced, as pointed out by Dumez and Minvielle (2022, p. 35), who indicate that "the mistake would be to see it only as a technological innovation, when it is a set of new uses in a given social context that is emerging". The rise of digital health, which aims to be ever more innovative and efficient, challenges us and leads us to a reflection that we present in section 13.2.2.

13.2.2. *Digital health in* smart cities*: opportunity or menace for meeting the challenges of 4P medicine*

The body of work focused on digital health and smart cities constitutes an opportunity to encourage practitioners, politicians and researchers to respond to and enrich the challenges and reflections related to 4P medicine (Flores et al. 2013; Avignon et al. 2015; Alonso et al. 2019). The concept of 4P medicine – predictive, personalized, preventive and participatory – was proposed in the early 2000s by Hood and Flores (2012). It involves an evolution of the description of symptoms and diseases treated under conventional curative medicine (standard approach to pathos) toward a medicine intended to rethink and transform the management of health, pathologies and medical practices (Chapel 2021).

For some years now, this 4P medicine has been promoted as a vital factor for the improvement of medical services (Alonso et al. 2019), where the challenges are significant: inventing strategies and technologies, managing the impact of this medicine on society, including the main ethical, social, legal, regulatory and economic problems, and making this medicine accessible to patients (Hood and Flores 2012).

At the crossroads of the development of smart cities, these challenges must be rigorously addressed by supervisory authorities and scientific organizations. In the next section, we propose a nuanced presentation of these challenges using a few examples.

First, genetic data from Big Data are captured and stored in the Cloud, researched, shared and analyzed. After this, the curative approach of traditional medicine is displaced by 4P medicine, which also includes an anticipatory (preventive and predictive) and individually participatory approach (Slim et al. 2021). Many applications, based on various features of AI and blockchain technology, are being explored to improve biomedical and healthcare data systems (Rajawat et al. 2022). These characteristics are diverse: decentralized management, immutable audits, authenticity of data, better security and even restoration of charterers' rights. They are currently a major source for scientific studies where some researchers envisage that they could replace randomized trials, which are often costly and difficult to implement (Rajawat et al. 2022). AI and the blockchain are thus used to create innovative solutions for improving current standards of processing, sharing, analysis and classification of medical data based on health outcomes in the field of smart cities (Treiblmaier et al. 2020; Du et al. 2021; Rajawat et al. 2022).

With this in mind, in November 2019, the Health Data Hub was created to enable the cross-referencing of public health databases and facilitate their use by numerous research and development teams in France. It is a one-stop shop that allows the optimization of artificial recognition and personalized prediction services based on AI. However, we must interpret the results of these initiatives with caution, as these subjects are still highly controversial. It has recently been shown that, depending on the method of collection, database results may lead to opposite results (Childers and Maggard-Gibbons 2021; Slim et al. 2021). Then, while access to Big Data comes from hospitals, research, community medicine or even connected objects, many voices (National Council of Bars, the National Order of Physicians or deputies) remain concerned about the hasty management of these projects. "Getting around the tree that hides the forest" is the suggestion of Bernard Fallery (2020), professor emeritus in information systems at the University of Montpellier. This means perceiving the extent of the questions raised by digital transformation in society in general and here in health. These questions are technical (centralization or interoperability of databases), legal (consent and medical confidentiality), economic (digital transformation through neo-liberalism) and social (toward a disciplinary and control society) (Fallery 2020).

The challenge in the coming years is to be able to understand all aspects of this personalized medicine by considering each person's uniqueness, but also that each individual be understood as a whole (Slim et al. 2021). To this end, due to the inherent complexity of the massive development of digital technology in health and to address all these organizational and ethical limitations, in the following section we develop the idea that smart cities require the collaboration of interdisciplinary teams before disseminating the concept of 4P medicine on a large scale.

13.3. Creation of solutions and initiatives in territories: when actors mobilize to innovate

13.3.1. *Interdisciplinary approach to chronic diseases*

While the world of health (in France) is presented as very institutionalized and compartmentalized (Laporte 2021), we consider that given human factors, institutional factors and the multidisciplinary dimension of the health sector, it is necessary to engage stakeholders in this movement to transform territories into smart and sustainable cities. Regardless of the use of ICT and the amount of data processed, to be smart, a city must integrate institutions, businesses, users and citizens (Desdemoustier and Crutzen 2015).

Among many studies devoted to the development of digital health solutions, some authors emphasize the interest of bringing together many stakeholders (developers, patients and health professionals) and maintaining an interdisciplinary team to enable dual expertise: content and technology (Matthew-Maich et al. 2016; Banbury et al. 2018; Xiong et al. 2018; Nussbaum et al. 2019). Notably, the literature on information systems has shown that the involvement and participation of these different stakeholders in information technology (IT) design projects for digital health are factors in the success of these projects (Habib et al. 2017). However, this participation is not self-evident, and the gap that may exist between the different actors who each have their own languages, logic and requirements may explain the failure of many projects (Granja et al. 2018; Baghizadeh et al. 2020).

To support these comments, we present an empirical study that was carried out within a project to design a mobile application in digital health: the e-health app project (Chapel et al. 2022). The aim was to analyze how project leaders constructed the legitimacy of this project to obtain the support of other stakeholders.

Indeed, if stakeholders are convinced that this application and the changes it implies will be beneficial to them, they will then offer their acceptance and support, which contributes to building the legitimacy of the project (Meissonier et al. 2015; Orlowski et al. 2016). Consequently, building legitimacy is important for gaining stakeholder participation and thus influencing the success or failure of this type of project (Kaganer et al. 2010; Flynn and Du 2012; Chapel et al. 2022).

Based on a single case study (Gagnon 2005), this research aims to analyze construction of the legitimacy of an IT design project in digital health. Data collection began in September 2018 and ended in January 2020. It represented more than 150 h of observation and 13 semi-structured interviews with all the actors among the two stakeholders studied in the project.

In Box 13.1, we explain the outlines of this project and the results of this study, which may represent an interesting example for understanding the first steps of the development of digital health in smart cities.

The e-health app project aims for the development and scientific validation of new functionalities and health services through the design of a mobile app and a digital site. Aimed at chronic respiratory diseases and the healthcare professionals who deal with them, this project integrates various connected objects and support for patients through the use of personalized programs of appropriate physical activities, nutrition and health education, as well as health monitoring by health professionals.

"We want to offer patients and healthcare professionals new means of support for chronic diseases through the principles of multidisciplinary rehabilitation" (doctor 1, project manager, semi-structured interview).

This multidisciplinary project assembles several stakeholders: a hospital-university team and a private health organization. The hospital-university team consists more precisely of a coordination team comprising two hospital doctors, the principals of the project and several researchers and teacher-researchers from three different disciplines, attached to four public research laboratories in different fields (management, psychology, physiology and computer engineering-production engineering).

Drawing on Suchman's theory of organizational legitimacy (Suchman 1995), this article identifies in particular the emergence of so-called strategies of legitimation maintenance or repair when project leaders detect a decrease in support for the project among the various stakeholders.

Below are two of the four cases identified.

First example: some teacher-researchers and health professionals have expressed their concern as to the means and skills available to enact the project in a comprehensive manner.

"In the end, I've never done this in my life, creating an app [...]. And it's actually very complicated to achieve what we want to do. We all have good ideas, but on the other hand we don't have a lot of resources. It's not financial resources. it's the resources of expertise" (doctor 5, semi-structured interview).

The project faced a decrease in cognitive legitimacy: difficulty in being "compliant with the standards" of IT design projects in health. The response of the project leaders was to implement a cognitive legitimation strategy: consulting expert opinion and thus encouraging information sharing with the help of "resource" people.

Second example: some teacher-researchers in the project mentioned the need to include patients more in decision-making. They showed their disagreement with the absence of patients among the stakeholders in project design.

"Ultimately, we are creating something for the user who is not even involved in the project" (teacher-researcher 1, plenary meeting).

The project leaders then faced another deficit of cognitive legitimacy: the unease of teacher-researchers associated with good professional practices in designing a mobile application that requires user involvement from the design stage.

To protect themselves from these unforeseen challenges of the project, project leaders sought to strengthen the legitimacy that they had already acquired. The strategy of "accumulating goodwill", derived from so-called moral legitimacy, was promoted for months.

Project leaders attempted to solve the problem by providing their goodwill (Suchman 1995) based on the actions suggested by collaborators to maintain their support.

Figure 13.2 represents the process of repairing the project's organizational legitimacy.

Box 13.1. *Challenges and strategies for legitimizing the e-health app project (Chapel et al. 2022)*

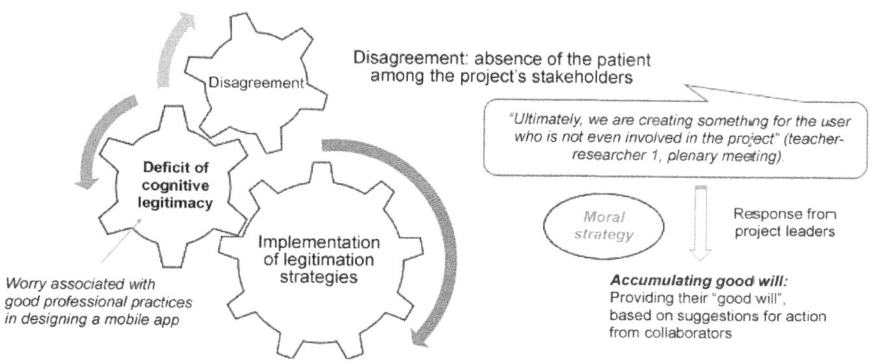

Figure 13.2. *Process of repairing the legitimacy of the e-health app project*

After every upheaval, we can see changes taking place, observe organizations evolving and watch how a system can transform itself (Orlikowski 1996; Tsutsui et al. 2021). The Covid-19 health crisis also widely enabled the acceleration of projects in the field of digital health.

So, in the next part of this chapter, we have decided to present another case in the field of telemedicine, a form of remote medical practice using ICT.

13.3.2. *Lockdown: an opportunity for better accessibility to care*

In telemedicine, we find two groups of applications: those that connect a patient with a health professional, such as home telemonitoring for the management of chronic diseases, and those that connect two health professionals, such as teleradiology (Saliba et al. 2012). Due to its characteristics, telemedicine is an important vector for improving access to care, particularly in fragile areas, to establish a diagnosis, ensure preventive or post-therapeutic monitoring, or even request a specialist opinion (Schwamm et al. 2009).

However, it should not replace the usual medical practices and should be complementary. In Box 13.2, we highlight the initiatives of a respiratory rehabilitation health network in the Occitania region, which combined these different practices during periods of lockdown.

> The AIR+R network, now known as Occitan'air, is a health network organized as a non-profit association under the 1901 law. Its aim is to ensure continuity of the care pathway for people suffering from chronic diseases with a predominantly respiratory component through a rehabilitation program. Rehabilitation is a comprehensive and individualized intervention that aims to improve health and promote adherence to good health behaviors, in particular through adapted physical activity (APA) and therapeutic patient education (Spruit et al. 2013; Rochester et al. 2015).
>
> During the lockdown, patients were deprived of their weekly APA session. The structure then allowed employees to train in supervising a remote session using a video conferencing application.
>
> How did this work, in practice? Patients possessing the appropriate equipment were able to benefit from the "tele-rehabilitation" program (use of ICT to provide rehabilitation services) (Kairy et al. 2009). A preliminary induction interview was organized with the practitioner to test the proper functioning of the tool and organize a practical space at home. Subsequently, patients could meet in groups of up to 12 in a virtual room to discuss and practice an adapted physical activity supervised by an APA teacher.

A study investigated the role of telerehabilitation in the transformation of patient pathways in this structure to enrich existing conceptualizations of the patient pathway (Chapel et al. 2021). In fact, management literature shows an interest in favoring more elaborate pathways rich in experiences (Jacob et al. 2021).

However, some practical limitations have already been raised within this health network, such as inequality of ability to understand and use digital technology (digital literacy), access to digital tools and internet connection at home. The telephone and motivational text messaging still represent solutions for overcoming this inequality.

And after lockdown? This tool was seen as an opportunity to develop new rehabilitation activities for chronically ill people.

This allowed the structure to become further engaged in this new project and make it a priority for 2022.

"It will allow us to offer our program to new people, to develop more therapeutic education workshops for patients already registered in our programs, to organize consultations or social meetings other than by telephone, and to offer training to professionals" (Hélène Forthin, responsible for AIR+R development, *newsletters* 30/1).

Box 13.2. *Accelerating digital engagement in the region: the case of the AIR+R health network*

13.4. Rethinking coordination of pathways across the territory: shared digital platforms

13.4.1. *Patient care pathways: coordination through information systems*

The patient pathway takes on meaning within the framework of comprehensive patient care and disrupts the traditional fragmentation of the healthcare sector. It consists of a succession of episodes of care provided by different professionals and whose complexity requires continual coordination (Pascal 2003; Véret and Bidan 2019; Lemaire et al. 2020). It is part of a transformation characterizing the French health system, marked by a succession of disruptions, including the Covid-19 pandemic, which aggravated the deep structural crisis experienced by health facilities (Kalika 2020). This concept of a pathway refers to a paradigm shift where the patient no longer has to adapt to the healthcare system, but rather the system must coordinate with the patient (Rodde-Dunet and Mounic 2016).

In this context, the use of information systems, and more broadly of digital health (or e-health), is set to introduce irreversible changes in the methods of healthcare production (Claveranne et al. 2013; Garidi 2021). By redefining information circuits within healthcare establishments and their environment, these technologies support coordination between the various stakeholders in the patient care pathway (Bonnardel and Zardet 2018). Digital health is thus propelled to an important role in the health sector, as shown by research in this field (Dumez et al. 2015; Ologeanu-Taddei and Paré 2017; Chapel 2021).

Nevertheless, several successive reports have noted a delay in the computerization of the health system with formulated recommendations, particularly when implementing systems of communication that promote exchanges between networked actors and converge around the patient's journey (Bertezene and Dubrion 2017; Sampieri-Teissier and Livolsi 2019).

The national strategy has translated into financial incentives provided by the government to accelerate the maturity of information systems within establishments by means of several plans, such as the digital hospital program, Hop'en and Sun-Es. These provide an opportunity for establishments to develop their computerized patient files, access and update the shared medical file, and use MSSanté messaging to share secure data between healthcare professionals and between professionals and patients, notably using the My Health Space system. This system, a space supported by health insurance and the Ministry of Health, aims to become a secure health record for each insured person.

Beyond the single aspect of information sharing, the aim of the digital strategies implemented is to place the patient at the center of a structured healthcare organization, for example, through territorial hospital groupings, which must have a convergent information system. The goal is to allow rationalization of human and financial resources to ensure the relevance of care within the territories, thus accentuating a need for coordination around the patient. In this sense, the relevance of patient care pathways in a territory constitutes both an objective and a means for the use of information systems.

At the heart of debates on the strategy for transforming the health system, relevance is defined by the High Authority for Health as: "The right care, at the right time, in the right place, for the right patient".

Thus, it is not limited to the scope of economy in gestures and actions, but leads healthcare establishments to optimize management of their patient flow with the aim of improving the efficiency of the resources mobilized while providing a response to patient needs (Garidi 2022). In this regard, coordination information systems have emerged, for example, ViaTrajectoire, which we will present in section 13.4.2.

13.4.2. *Coordination system for pathway relevance*

ViaTrajectoire

Viatrajectoire is software from the SARA Health Cooperation Group, initially financed by the Rhône-Alpes Regional Health Agency. Hosting and operation of the software is supported by the Hospices Civils de Lyon.

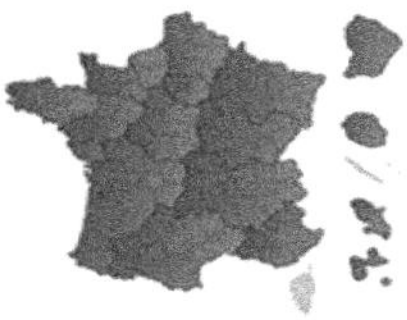

17 partner regions
145,821 declared professionals
41,433 private doctors
515,356 elderly people and family carers

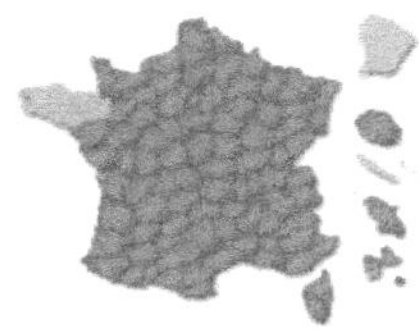

4,933 healthcare establishments
7,353,075 healthcare records

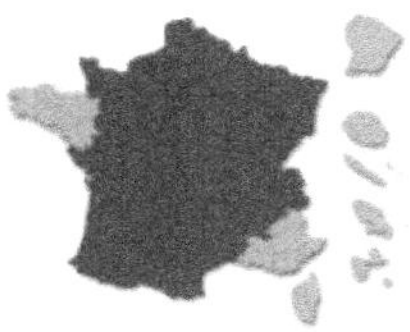

9,544 establishments for elderly people
1,063,890 records of elderly people

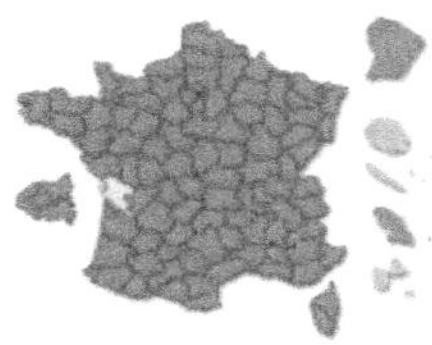

12,186 establishments for people with disabilities
1,093,203 records of people with disabilities
2,183,183 guidance decisions for people with disabilities

This tool has been deployed gradually over 17 regions. We present the experience of a follow-up care and rehabilitation establishment located in department 57 - Moselle, whose deployment was announced by the ARS Grand Est in December 2020. This was implemented in several stages from January to May 2021 ranging from configuring the tool to training professionals to using the tool. Table 1 below presents some figures on the admission files processed by the establishment. They reflect an analysis of the requests received to avoid unsuitable pathways in SSR.

Total Requests received from 01/06/2021 to 01/06/2022	Prescription followed by an admission		Requests followed by a cancellation:				
	Effective admissions in the structure studied	*Effective admissions outside the structure studied (cancelled exchanged)*	**Cancelled requests**	*Requests cancelled without response from the structure studied*	*Requests cancelled after response from the structure studied*	*Requests cancelled following at least one approval from the structure studied*	*Requests cancelled following at least one refusal from the structure studied*
171	70	27	74	9	65	5	62

Figure 13.3. *Presentation of an overview of the deployment of ViaTrajectoire. For a color version of all figures in this chapter, see www.iste.co.uk/ ngoala/smartcity.zip*

Establishment of a patient care pathway positions the information system as a coordination tool that integrates all the stakeholders involved. To illustrate this point, we present ViaTrajectoire, a public information system, accessible free of charge and secure, which offers personalized guidance assistance as part of a care pathway that requires admission to health establishments, establishments for the elderly and establishments for people with disabilities.

Access to this tool is granted to professionals, but also to patients, elderly people and family caregivers. It thus offers a detailed directory of the resources available in a territory to help with orientation when faced with multiple and varied options. Despite the delay in digital technology indicated by the Ségur de la santé in 2020, we are witnessing an irreversible digital transformation through these systems that encourage rethinking of the coordination of pathways.

Furthermore, these new information systems are changing the historically paternalistic patient–doctor relationship, where the doctor has knowledge and a prescription and the patient carries it out, toward a paradigm shift, that of the patient-expert (Pierron 2007). This paradigm shift is reinforced by a digital health democracy movement, giving rise to the "e-patient", an actor engaged in their own care, who has given themselves the means to act through information systems (Ferguson and Frydman 2004; Dumez and Minvielle 2017).

Despite the certain interest generated by these new information systems, they are confronted with territories marked by a strong digital disparity and a need for sharing in the interests of relevant orientation of patient care. This complexity includes the uniqueness of each patient's profile and the variety of care processes (Pascal 2000; Minvielle 2021). Also, these coordination information systems are still, for the most part, designed by professionals for the patient rather than with the patient (Bélanger et al. 2012), although engagement from patients contributes to pathway relevance by reducing unjustified hospitalizations (Mockford et al. 2012; Richards et al. 2013). ViaTrajectoire embodies a fundamental change in ways of involving the patient in their own care despite the technical and organizational difficulties encountered in deploying this tool. Patient engagement also relies on access to ICT (as shown in the report on the AIR+R network), and more broadly on digital health. However, in France, Baena and Rachiq (2018) reveal that 14 million French people, or 28% of the population, are estranged from digital technology, which constitutes a real obstacle to patient engagement, given the context of an increasing digitalization of activities.

13.5. Conclusion

The use of digital health is growing and becoming essential for smart cities. Increasingly present in the daily lives of patients, it nevertheless requires a significant need for information and support.

Clearly, in connected cities, digital health alone cannot be the miracle solution for modernizing the healthcare system in France, just as it is not the panacea that will solve the problem of medical deserts. However, if it is made one of the strategic axes of the care pathway, it should be able to contribute effectively due to the technological facilitations that it allows, provided that it can benefit from a willingness on the part of professionals themselves to work in synergy and collaboration toward better patient care. To do this, digital health needs strong leaders, particularly medical ones, to implement this difficult change management, overcome resistance and develop medical practices based on an inherited form of organization which is now a liability.

Although the results of health initiatives in smart cities are still mixed, they nonetheless have positive consequences for the health of citizens and offer real opportunities to adapt health systems. Many scientific, training and territorial projects are in development and the massive processing of health data in a Big Data approach should promote the development of personalized medicine.

But beyond the purely technical and/or organizational aspects linked to the deployment of digital health, beyond the advantages that it indicates in terms of reorganizing the healthcare pathway and combating inequalities in access to care, digital health will also have to meet the challenge of full recognition and thus of questioning of certain developments, notably its financing by health insurance. There remains another challenge, and not the least, that of concretizing its legal framework in terms of:

– respect for confidentiality and the necessary security of patient data;

– obtaining informed consent from patients with regard to this new practice;

– delineation of individual and/or shared responsibilities between health professionals.

The challenge is not informational or technological; it is above all social, ethical and economic. It must absolutely be subject to strategic policy concerning its deployment at both regional and national levels, to allow the emergence and realization of a genuine economic sector, as visible as that successfully developed by France in other areas (rail, aeronautics, automobile, etc.).

13.6. References

Alam, T. (2021). Blockchain cities: The futuristic cities driven by Blockchain, big data and Internet of Things. *GeoJournal*, 1–30.

Allam, Z., Tegally, H., Thondoo, M. (2019). Redefining the use of big data in urban health for increased liveability in smart cities. *Smart Cities*, 2(2), 259–268.

Alonso, S.G., de la Torre Diez, I., Zapirain, B.G. (2019). Predictive, personalized, preventive and participatory (4P) medicine applied to telemedicine and eHealth in the literature. *Journal of Medical Systems*, 43(5), 1–10.

Avignon, A., Attalin, V., Boegner, C., Picandet, M., Sultan, A. (2015). Maladies chroniques non transmissibles : la solution est-elle numérique ? Partie 1. Chronicité et complexité. *Médecine des Maladies Métaboliques*, 9(2), 227–232.

Baena, A. and Rachiq, C. (2018). *Les bénéfices d'une meilleure autonomie numérique*. France Stratégie, Paris.

Baghizadeh, Z., Cecez-Kecmanovic, D., Schlagwein, D. (2020). Review and critique of the information systems development project failure literature: An argument for exploring information systems development project distress. *Journal of Information Technology*, 35(2), 123–142.

Banbury, A., Nancarrow, S., Dart, J., Gray, L., Parkinson, L. (2018). Telehealth interventions delivering home-based support group videoconferencing: Systematic review. *Journal of Medical Internet Research*, 20(2), e25.

Bélanger, E., Bartlett, G., Dawes, M., Rodríguez, C., Hasson-Gidoni, I. (2012). Examining the evidence of the impact of health information technology in primary care: An argument for participatory research with health professionals and patients. *International Journal of Medical Informatics*, 81(10), 654–661.

Bertezene, S. and Dubrion, B. (2017). La coopération dans les réseaux de santé en gérontologie. *La revue des sciences de gestion*, 1, 73–84.

Bonnardel, M. and Zardet, V. (2018). Cahier : les mutations socio-économiques des acteurs de la santé. *Revue Management & Avenir*, 100, 135–142.

Chapel, B. (2021). Les enjeux d'une intervention multidimensionnelle dans les maladies chroniques : de la réhabilitation à la téléréhabiliation. PhD Thesis, Université de Montpellier, Montpellier.

Chapel, B., Cases, A.-S., Ologeanu-Taddei, R., Hayot, M. (2021). Explorer le parcours patient à travers l'expérience patient : méthode mixte dans la téléréhabilitation (design d'étude). In *9ème colloque Santé*, Marseille.

Chapel, B., Ologeanu-Taddei, R., Cases, A.-S., Hayot, M. (2022). Défis et stratégies de légitimation d'une organisation temporaire : étude de cas d'un projet de conception d'une application mobile en e-santé. *Systemes d'information management*, 27(1), 67–99.

Childers, C.P. and Maggard-Gibbons, M. (2021). Same data, opposite results? A call to improve surgical database research. *JAMA Surgery*, 156(3), 219–220.

Claveranne, J.P., Pascal, C., Mick, S. (2013). L'innovation ou l'apocalypse. *Management Avenir Sante*, 1, 5–10.

Delmotte, H. (2015). "De la e-santé à la santé connectée" : accompagner la transformation de notre modèle de santé. *Hegel*, 2, 123–126.

Desdemoustier, J. and Crutzen, N. (2015). Smart Cities en Belgique : analyse qualitative de 11 projets. Report, HEC-ULg.

Du, X., Chen, B., Ma, M., Zhang, Y. (2021). Research on the application of blockchain in smart healthcare: Constructing a hierarchical framework. *Journal of Healthcare Engineering*.

Dumez, H. and Minvielle, É. (2017). L'e-santé rend-elle la démocratie sanitaire pleinement performative ? *Systèmes d'information management*, 22(1), 9–37.

Fallery, B. (2020). Données de santé : l'arbre StopCovid qui cache la forêt Helath Data Hub. *The Conversation*.

Ferguson, T. and Frydman, G. (2004). The first generation of e-patients, the new medical colleagues could provide sustainable healthcare solutions. *British Medical Journal*, 15, 1148–1149.

Flores, M., Glusman, G., Brogaard, K., Price, N.D., Hood, L. (2013). P4 medicine: How systems medicine will transform the healthcare sector and society. *Personalized Medicine*, 10(6), 565–576.

Flynn, D. and Du, Y. (2012). A case study of the legitimation process undertaken to gain support for an information system in a Chinese university. *European Journal of Information Systems*, 21(3), 212–228.

Gagnon, Y.-C. (2005). *L'étude de cas comme méthode de recherche : guide de réalisation*. Les Presses de l'Université du Québec, Montreal.

Garidi, S. (2021). L'impact des systèmes d'information hospitaliers sur la sécurisation du parcours de prise en charge du patient. PhD Thesis, Université de Lorraine, Nancy.

Granja, C., Janssen, W., Johansen, M.A. (2018). Factors determining the success and failure of eHealth interventions: Systematic review of the literature. *Journal of Medical Internet Research*, 20(5).

Green, S. and Vogt, H. (2016). Personalizing medicine in silico and in socio. *Humana. Mente Journal of Philosophical Studies*.

Grenier, C. and Oiry, E. (2021). *Altering Frontiers: Organizational Innovations in Healthcare*. ISTE Ltd, London, and John Wiley & Sons, New York.

Habib, J., Béjean, M., Dumond, J.-P. (2017). Grasping the transformational dynamics of digital health by integrating the perceptions of actors. *Systèmes d'Information et Management*, 22(2).

Hood, L. and Flores, M. (2012). A personal view on systems medicine and the emergence of proactive P4 medicine: Predictive, preventive, personalized and participatory. *New Biotechnology*, 29(6), 613–624.

Jacob, F., Pez, V., Volle, P. (2021). Principes, modalités, apports et limites de la Design Science Research en marketing-application illustrative au pilotage des parcours clients. *Recherche et Applications en Marketing*, 1–30.

Kaganer, E.A., Pawlowski, S.D., Wiley-Patton, S. (2010). Building legitimacy for IT innovations: The case of computerized physician order entry systems. *Journal of the Association for Information Systems*, 11(1), 2.

Kairy, D., Lehoux, P., Vincent, C., Visintin, M. (2009). A systematic review of clinical outcomes, clinical process, healthcare utilization and costs associated with telerehabilitation. *Disability and Rehabilitation*, 31(6), 427–447.

Kalika, M. (2020). *L'impact de la crise sur le management*. Edition Management et Société, Caen.

Kalkman, S., van Delden, J., Banerjee, A., Tyl, B., Mostert, M., van Thiel, G. (2022). Patients' and public views and attitudes towards the sharing of health data for research: A narrative review of the empirical evidence. *Journal of Medical Ethics*, 48(1), 3–13.

Lemaire, N., Polton, D., Tajahmady, A. (2020). Article 51 : expérimenter pour innover mais aussi innover dans la façon d'expérimenter. *Les Tribunes de la Santé*, 63(1), 35–42.

Lindenmeyer, C. and d'Ortho, M.P. (2022). *Santé connectée*. CNRS Éditions via OpenEdition, Paris.

Matthew-Maich, N., Harris, L., Ploeg, J., Markle-Reid, M., Valaitis, R., Ibrahim, S., Gafni, A., Isaacs, S. (2016). Designing, implementing, and evaluating mobile health technologies for managing chronic conditions in older adults: A scoping review. *JMIR mHealth and uHealth*, 4(2).

Meissonier, R., Lapointe, L., Houzé, E. (2015). Légitimer l'implémentation d'un ERP : une comparaison Canada-Thailande. *Journal of Decision Systems*, 24(3), 318–338.

Millard, N. (2020). How South Korea's smart city startups curbed the spread of COVID-19. *e27* [Online]. Available at: https://e27.co/how-south-koreas-smart-city-startups-curbed-the-spread-of-covid-19-20200323/.

Minvielle, É. (2018). *Le patient et le système : en quête d'une organisation surmesure : approches innovantes du parcours de santé*. Seli Arslan, Paris.

Mockford, K. and O'Grady, H. (2017). Prevention of surgical site infections. *Surgery (Oxford)*, 35(9), 495–499.

Mouton, M., Ducey, A., Green, J., Hardcastle, L., Hoffman, S., Leslie, M., Rock, M. (2019). Towards "smart cities" as "healthy cities": Health equity in a digital age. *Canadian Journal of Public Health*, 110(3), 331–334.

Nussbaum, R., Kelly, C., Quinby, E., Mac, A., Parmanto, B., Dicianno, B.E. (2019). Systematic review of mobile health applications in rehabilitation. *Archives of Physical Medicine and Rehabilitation*, 100(1), 115–127.

Ologeanu-Taddei, R. and Paré, G. (2017). Technologies de l'information en santé : un regard innovant et pragmatique. *Systèmes d'information management*, 22(1), 3–8.

Orlikowski, W.J. (1996). Improvising organizational transformation over time: A situated change perspective. *Information Systems Research*, 7(1), 63–92.

Orlowski, S., Lawn, S., Matthews, B., Venning, A., Wyld, K., Jones, G., Winsall, M., Antezana, G., Schrader, G., Bidargaddi, N. (2016). The promise and the reality: A mental health workforce perspective on technology-enhanced youth mental health service delivery. *BMC Health Services Research*, 16(1), 1–12.

Pascal, C. (2003). La gestion par processus à l'hôpital entre procédure et création de valeur. *Revue française de gestion*, 5, 191–204.

Payton, F.C., Pare, G., Le Rouge, C.M., Reddy, M. (2011). Health care IT: Process, people, patients and interdisciplinary considerations. *Journal of the Association for Information Systems*, 12(2), 4.

Pierron, J.P. (2007). Une nouvelle figure du patient ? Les transformations contemporaines de la relation de soins. *Sciences sociales et santé*, 25(2), 43–66.

Pulciani, S., Di Lonardo, A., Fagnani, C., Taruscio, D. (2017). P4 medicine versus hippocrates. *Annali dell'Istituto Superiore di Sanità*, 53(3), 185–191.

Rajawat, A.S., Bedi, P., Goyal, S., Shaw, R.N., Ghosh, A., Aggarwal, S. (2022). AI and blockchain for healthcare data security in smart cities. In *AI and IoT for Smart City Applications*, Piuri, V., Shaw, R.N., Ghosh, A., Islam, R. (eds). Springer, Singapore, 185–198.

Rathee, D.S., Ahuja, K., Hailu, T. (2019). Role of electronics devices for E-health in smart cities. In *Driving the Development, Management, and Sustainability of Cognitive Cities*, Ahuja, K. and Khosla, A. (eds). IGI Global, Hershey, PA, 212–233.

Richards, T., Montori, V.M., Godlee, F., Lapsley, P., Paul, D. (2013). Let the patient revolution begin. *British Medical Journal*, 346, 2614.

Rochester, C.L., Vogiatzis, I., Holland, A.E., Lareau, S.C., Marciniuk, D.D., Puhan, M.A., Spruit, M.A., Masefield, S., Casaburi, R., Clini, E.M. et al. (2015). An official American Thoracic Society/European Respiratory Society policy statement: Enhancing implementation, use, and delivery of pulmonary rehabilitation. *American Journal of Respiratory and Critical Care Medicine*, 192(11), 1373–1386.

Rodde-Dunet, M.H. and Mounic, V. (2016). Le parcours du patient selon la Haute Autorité de santé. *Journal des Maladies Vasculaires*, 41(2), 100.

Saliba, V., Legido-Quigley, H., Hallik, R., Aaviksoo, A., Car, J., McKee, M. (2012). Telemedicine across borders: A systematic review of factors that hinder or support implementation. *International Journal of Medical Informatics*, 81(12), 793–809.

Sampieri-Teissier, N. and Livolsi, L. (2019). L'hôpital dans la supply chain santé : quelle évolution des compétences managériales ? *Logistique & Management*, 27(1), 33–43.

Schwamm, L.H., Holloway, R.G., Amarenco, P., Audebert, H.J., Bakas, T., Chumbler, N.R., Handschu, R., Jauch, E.C., Knight, W.A. IIII, Levine, S.R. et al. (2009). A review of the evidence for the use of telemedicine within stroke systems of care: A scientific statement from the American Heart Association/American Stroke Association. *Stroke*, 40(7), 2616–2634.

Slim, K., Selvy, M., Veziant, J. (2021). Innovation conceptuelle : la médecine 4P et la chirurgie 4P. *Journal de Chirurgie Viscérale*, 158(3), S13–S18.

Solanas, A., Patsakis, C., Conti, M., Vlachos, I.S., Ramos, V., Falcone, F., Postolache, O., Pérez-Martinez, P.A., Di Pietro, R., Perrea, D.N., Ballesté, A.M. (2014). Smart health: A context-aware health paradigm within smart cities. *IEEE Communications Magazine*, 52(8), 74–81.

Spruit, M.A., Singh, S.J., Garvey, C., ZuWallack, R., Nici, L., Rochester, C., Hill, K., Holland, A.E., Lareau, S.C., Man, W.D.-C. et al. (2013). An official American Thoracic Society/European Respiratory Society statement: Key concepts and advances in pulmonary rehabilitation. *American Journal of Respiratory and Critical Care Medicine*, 188(8), e13–e64.

Treiblmaier, H., Rejeb, A., Strebinger, A. (2020). Blockchain as a driver for smart city development: Application fields and a comprehensive research agenda. *Smart Cities*, 3(3), 853–872.

Tsutsui, M., Gerayeli, F., Sin, D.D. (2021). Pulmonary rehabilitation in a post-COVID-19 world: Telerehabilitation as a new standard in patients with COPD. *International Journal of Chronic Obstructive Pulmonary Disease*, 16, 379.

Véret, M. and Bidan, M. (2019). Patient Chain Management (PCM) – Patient Pathway Management (PPM) : une approche organisationnelle multidimensionnelle. In *Supply Chain Management de la santé*, Bentahar, O. and Benzidia, S. (eds). EMS Éditions, Caen.

Xiong, S., Berkhouse, H., Schooler, M., Pu, W., Sun, A., Gong, E., Yan, L.L. (2018). Effectiveness of mHealth interventions in improving medication adherence among people with hypertension: A systematic review. *Current Hypertension Reports*, 20(10), 86.

14

Learning

14.1. Introduction: creating new learning experiences accessible to all

Among the different domains of the smart city, education refers to the smart people who make up the city (Anthopoulos 2017). Without their participation, the smart city does not exist. For Caragliu et al. (2011), the level of urban intelligence depends on the education of the citizens and users of the city. On the one hand, the presence of creative citizens favors co-construction of the smart city based on their greater involvement in governance (Hudson et al. 2019), but also in urban experiments (Snow et al. 2016).

On the other hand, the ability of users to accept technology is a key factor in the successful development of smart cities that use digital technology to provide new smart services.

Indeed, many smart services today, whether private or public, aim to support users in adopting more responsible practices according to their level of education such as the adoption of autonomous vehicles (Park et al. 2021) or the acceptability of "smart homes" (Baudier et al. 2020).

Smart cities are also attractive to users because they concentrate many professional opportunities, or opportunities for training, apprenticeship, etc., that are qualified and better paid. Winters (2011) shows that smart cities are often "higher education centers" to which students move and, once they graduate, stay there to live and work.

Indeed, many digital tools and a wide variety of content now favor the creation of rich learning experiences to further engage learners in class or remotely, from primary to professional training, including higher education. With access to these new

---

Chapter written by Sara LAURENT.

experiences, new threats risk fostering a divide between learners, institutions, but also territories. It is thus necessary to reflect on the design of these new learning experiences: how and why should we rethink the design of learning experiences? What are the challenges for learners, but also for the institutions and teachers who create these new designs?

14.2. Designing new learning experiences

With the evolution of learners' digital uses and practices, the education and training sectors are constantly evolving to integrate new tools and content for enriching the learning experience.

14.2.1. *A variety of tools for an enhanced learning experience*

According to Baron's (2014) history, the ubiquity of educational resources first played a major role in the digital transformation of education. The teacher is no longer the only "medium" that delivers knowledge; the learner has access to a multitude of sources. Then came the trend toward individualization of teaching based on a better understanding of learners and their level of learning. Finally, with the sophistication of information and communication technologies, the use of blogs and social media has promoted a community learning experience.

Thus, whether at primary or secondary school, in higher education, or in professional training, the learning experience can be enriched by a diversity of tools and content, which may be:

– internal to the training institution, or external (social networks and blogs, online video platforms, etc.);

– digital, with a more or less sophisticated level of technology (online questionnaires, social robots for telepresence, blockchain to verify diplomas, etc.);

– intended for use before, during and after class or training;

– before class/training to prepare the experience, create the content;

– during class/training, the tools and content must encourage learners' engagement with new ways of participating, making them actors in their own learning;

– after class/training, to assess the progress of learners, at the group and institutional level, and to best adapt the experience, tools, and content being used.

Furthermore, the ever-increasing rate of learner ownership of devices promotes accessibility to inclusive, equitable and sustainable education (Burbules et al. 2020).

This represents an opportunity for the development of "lifelong" training to better support learners' employability through iterative learning.

14.2.2. *Motivations to adopt digital technologies for designing new learning experiences*

Molnar (2019) discusses the different reasons motivating the adoption of technology in education: (1) drawing from a greater diversity of low-cost content and media; (2) for a richer learning experience ; and (3) interactive and personalized for learners.

1) More content: learners' attention span decreases every twenty minutes. Educational sciences recommend designing courses and training in 20-minute sessions to stimulate learning: activity, video, debate, group work, etc. Many digital tools and devices now exist to facilitate and manage class time, but also outside the classroom, with so-called asynchronous activities. Thus, students can access a large number of educational resources via a platform (e-learning), choose an online course (MOOC), etc. Although this requires additional effort to classroom learning, this mixed approach encourages engagement by learners who co-construct their courses and their training (Means et al. 2013).

2) More interaction: interactions with teachers and trainers, but also with other learners, are facilitated by digital technology. Several channels are possible: email, social media, online chats, etc. Learners no longer learn only in the classroom, but also outside the classroom, with other learners or other experts all over the world (Zhao and Watterston 2021). The roles of teachers and trainers are evolving: they no longer just deliver knowledge, but also support learners in interacting and co-constructing their learning experience.

3) More personalization: digital tools and devices can also track learners' progress, difficulties and achievements. According to Krishnamurthy (2020), in the case of higher education, data collected from student assessments should help identify learning needs and provide the most personalized experience possible (choice of the most engaging content and media for each student, with the highest success rates in exams, etc.). For Tlili et al. (2023), the learning style must evolve based on the learner's personality.

In September 2020, the NEOMA Business School launched its virtual campus to welcome, in the form of avatars, its students, professors and administrative staff, that is, nearly 10,000 people spread over three geographical areas in France: Reims, Rouen and Paris. Since 2018, the school has explored and used a range of technological tools to address a problem: how to create a shared educational experience for students who may be geographically far apart? Alain Goudey, Director General in charge of digital technology at the school, discusses the educational challenges linked to this innovation for enriching the student experience, in addition to the digital tools already deployed.

"In terms of experience, students did not necessarily expect to have an educational experience on this type of platform. There is a surprising, astonishing, even 'bizarre' side for some of them. On the other hand, they all recognize that the experience is extremely positive. Once you get past the surprise, 'hey, this looks like a game', you actually manage to create an educational experience that is extremely interesting because it completely hybridizes the real world on campus, on site, with digital platforms that use videoconferencing tools, which are too 'flat'.

What is quite interesting for students and teachers is that the virtual campus allows them to rediscover 'pedagogical dramaturgy'. Pedagogical dramaturgy involves a unity of place, a unity of action and a unity of time. Of course, we find ourselves in a virtual environment, but you have this unity of place which teachers and students share through their avatar.

And when we look at the theoretical elements that explain the success of this approach, two major theories are interesting. The first is the theory of the extended self, that is, when we create our avatar, we extend the self. So, we are not necessarily going to create an exact copy of ourselves. The avatar is a form of extension of one's personality, one's being and one's individuality. And the second theory is that of embodied cognition, which says that all cognitive activity is sensorimotor in nature, including high-level activities, whether it be language learning or training. This explains why being represented in the form of an avatar makes perfect sense in an activity of the educational experience type, since we are recreating something of a sensory-motor nature. It is richer in this type of environment than using '2D' videoconferencing tools.

There is also more engagement and more spontaneity. From the moment that we create a persistent virtual world, we recreate spontaneity and we recreate nonlinearity in exchanges. When you do a day of classes by videoconference: you have your meeting from 9 a.m. to 10 a.m., then the next one from 10 a.m. to 11 a.m., etc., and in the end, like this, you are chaining timeslots together in a linear and mechanical way. Running after the clock, more or less, without necessarily being able to take breaks. When you are on a platform, such as our virtual campus, you are recreating the unexpected and the nonlinear. Although you may have scheduled a class, for example, a 3-h session, the teacher can schedule a break or two and do so whenever they want. And when break time comes, not everyone turns off their computer and goes to do something else. Often, a large majority of people take their avatar, wander off elsewhere on campus and chat informally. I have regularly observed, at NEOMA events on the virtual campus, that students who had not seen each other for several months because of the coronavirus met up 'by chance', at least without having planned it, inside this persistent world, and began to exchange and discuss".

Box 14.1. *Testimony of Alain Goudey, Director General Director in charge of digital technology at NEOMA Business School*

14.3. Thinking about the accessibility of new learning experiences

There must be as many learning experiences as there are learners and different institutions. Depending on the rate of device ownership and their digital skills, several points of vigilance must be considered to reduce the risk(s) of a digital divide within the educational ecosystem.

14.3.1. *At learner level*

Although learners are better equipped (smartphones, laptops and tablets), they do not all make optimal use of all of the functionalities offered by these increasingly sophisticated tools. On a daily basis, learners communicate easily over social media, create and share photo and video content, but for their learning experience, other practices come into play. For Ivanenko and Artamonova (2020), this means distinguishing between learners' digital skills and their digital competence. Digital skills refer to the technical skills for using digital technologies; while digital competence is a broader concept that includes technical skills, but also the ability to adapt usage to the environment (work, studies, leisure), and the ability to question this usage and the motivation to participate in digital culture (Ilomäki 2011). For all users of digital technologies, whether learners or otherwise, digital ease is a major issue of public interest to enable as many as possible to participate in an increasingly connected environment (Portes 2018; Cauchard and N'Goala 2021).

In addition, the new tools for delivering these experiences raise questions about the reliability of their security for storing and managing learner data (financial and social information, assessments, verification of diplomas, etc.). The complexity of the different technologies should however not make management of learner data opaque. The data stored by educational and training establishments must be easy to consult to avoid potential abuses in learner profiling and the introduction of selection bias. Treiblmaier et al. (2020) encourage the use of solutions based on the blockchain for security and traceability of data linked to learner files.

14.3.2. *Within institutions and among teachers*

The heterogeneity that exists among learners creates more difficulties for teachers and trainers who must equip themselves and be able to *design* these new learning experiences (Béduchaud et al. 2020). With the health crisis linked to Covid-19, teachers have used digital technology without any preparation or adaptation time to create new online learning experiences. Prior to this crisis, Wingo et al. (2017) had already studied perceived barriers to online teaching. Teachers were concerned about students' ability to successfully follow online learning, the lack of technical support and the risk of additional workload in developing this online teaching. In the wake of the health crisis, several studies have reported on the challenges faced by teachers in continuing their activity online. First, on a logistical level, the sudden change left no time for institutions to properly equip their teaching staff: poor Internet access and lack of equipment (Sari and Nayir 2020). For Marinoni et al. (2020), this crisis should encourage accelerated dematerialization of educational resources and evaluation methods, and the digitalization of administrative procedures. Second, on a human level, teachers were not all equal in terms of digital skills and online teaching to create suitable learning experiences, create connections and manage a class (Sari and Nayir 2020; Sá et al. 2021). In this context, teachers are also learners.

The results of the Teaching and Learning International Survey (TALIS), conducted and published by the OECD in April 2022, revealed that beyond the availability of digital equipment, the disparities in digital learning between schools worldwide depend heavily on the distribution of teachers. Teachers with a strong sense of personal efficacy in using digital technology are mainly present in private or socioeconomically advantaged institutions, because these latter have up-to-date digital tools (OECD 2022). Thus, learner accessibility to new learning experiences does not depend on the simple presence of digital tools, but also on learners' ability to appropriate these new experiences, and on the ability of teachers to support them in this appropriation.

14.4. Conclusion: challenges linked to education at the territorial level

In the smart city, education is a lever for citizen participation in the governance of the city and the pursuit of sustainable development goals. Education involves, among other things, the presence of educational and training institutions aimed at learners of all ages. Between physical and digital environments, the smart city offers a multitude of resources (material, human, economic, etc.) to educational and training institutions for designing new learning experiences.

However, the distribution of these resources leads to a disparity between learners who have different abilities, between institutions, but also between territories. At the societal level, the smart city, as a "center of higher education", must ensure that it provides access to education for all, beyond its borders (Winters 2011). In fact, the smart city, through concentration of material and human resources, may promote exclusion of users who are less comfortable with digital technologies (Borkowska and Osbourne 2018). According to Eppley and Shannon (2015), with the high concentration of smart services within cities, a gap is widening between rural and urban areas. The smart city must therefore be sufficiently connected to the territories to make education and the appropriation of digital technologies accessible, and thus allowing more citizens to participate in collective intelligence, even outside of the cities.

14.5. References

Anthopoulos, L.G. (2017). *Understanding Smart Cities: A Tool for Smart Government or an Industrial Trick?* Springer, Cham.

Baron, G.L. (2014). Elèves, apprentissages et “numérique” : regard rétrospectif et perspectives. *Recherches en éducation*, 18, 97–99.

Baudier, P., Ammi, C., Deboeuf-Rouchon, M. (2020). Smart home: Highly-educated students’ acceptance. *Technological Forecasting and Social Change*, 153, 119355.

Béduchaud, D., Coudray, A., Coureau-Falquerho, E. (2020). *École, numérique et confinement : quels sont les premiers résultats de la recherche en France*. Éditions Canopé, Paris.

Borkowska, K. and Osborne, M. (2018). Locating the fourth helix: Rethinking the role of civil society in developing smart learning cities. *International Review of Education*, 64(3), 355–372.

Burbules, N.C., Fan, G., Repp, P. (2020). Five trends of education and technology in a sustainable future. *Geography and Sustainability*, 1(2), 93–97.

Caragliu, A., Del Bo, C., Nijkamp, P. (2009). Smart cities in Europe. PhD Thesis, VU University Amsterdam, Amsterdam.

Cauchard, L. and N’Goala, G. (2021). L’inclusion numérique : se rapprocher de l’usager pour des stratégies plus efficaces. In *Impacts sociétaux et développement du management des technologies organisationnelles*, Mocquet, B., Pélissier, C., Cotreaux, S. (eds). Presses des Mines, Paris.

Eppley, K. and Shannon, P. (2015). Literacy education for the Lumps and Divots of smart cities and rural places. In *Educational Opportunity in Rural Contexts: The Politics of Place*, Williams, S.M. and Grooms, A.A. (eds). Information Age Publishing, Charlotte, NC.

Hudson, L., Wolff, A., Gooch, D., Van Der Linden, J., Kortuem, G., Petre, M., O’Connor-Gotra, S. (2019). Supporting urban change: Using a MOOC to facilitate attitudinal learning and participation in smart cities. *Computers & Education*, 129, 37–47.

Ilomäki, L., Kantosalo, A., Lakkala, M. (2011). *What is Digital Competence?* European Schoolnet, Brussels.

Ivanenko, M.N. and Artamonova, M.V. (2020). Employees’ competitiveness in digital transformation. In *International Scientific Conference Far East Con* (ISCFEC 2020). Atlantis Press, Amsterdam.

Krishnamurthy, S. (2020). The future of business education: A commentary in the shadow of the Covid-19 pandemic. *Journal of Business Research*, 117, 1–5.

Marinoni, G., Van't Land, H., Jensen, T. (2020). The impact of Covid-19 on higher education around the world. Global Report, IAU.

Means, B., Toyama, Y., Murphy, R., Baki, M. (2013). The effectiveness of online and blended learning: A meta-analysis of the empirical literature. *Teachers College Record*, 115(3), 1–47.

Molnar, A. (2019). Education in smart cities: Practices and challenges. In *Proceedings of the IADIS International Conference Applied Computing*, Utrecht, 11–13.

Molnar, A. (2021). Smart cities education: An insight into existing drawbacks. *Telematics and Informatics*, 57, 101509.

OECD (2022). *Comment rendre l'accès des élèves à l'apprentissage numérique plus équitable ?* OECD Publishing, Paris. doi: 10.1787/fdf2470c-fr.

Park, J., Hong, E., Le, H.T. (2021). Adopting autonomous vehicles: The moderating effects of demographic variables. *Journal of Retailing and Consumer Services*, 63, 102687.

Portes, A. (2018). La transparence numérique : rôle du client et conséquences sur la relation à la marque. PhD Thesis, Université Montpellier, Montpellier.

Sá, M.J., Santos, A.I., Serpa, S., Miguel Ferreira, C. (2021). Digitainability–digital competences post-COVID-19 for a sustainable society. *Sustainability*, 13(17), 9564.

Sari, T. and Nayır, F. (2020). Challenges in distance education during the (Covid-19) pandemic period. *Qualitative Research in Education*, 9(3), 328–360.

Snow, C.C., Håkonsson, D.D., Obel, B. (2016). A smart city is a collaborative community: Lessons from smart Aarhus. *California Management Review*, 59(1), 92–108.

Tlili, A., Denden, M., Essalmi, F., Jemni, M., Chang, M., Kinshuk, Chen, N.S. (2023). Automatic modeling learner's personality using learning analytics approach in an intelligent Moodle learning platform. *Interactive Learning Environments*, 31(5), 2529–2543.

Treiblmaier, H., Rejeb, A., Strebinger, A. (2020). Blockchain as a driver for smart city development: Application fields and a comprehensive research agenda. *Smart Cities*, 3(3), 853–872.

Wingo, N.P., Ivankova, N.V., Moss, J.A. (2017). Faculty perceptions about teaching online: Exploring the literature using the technology acceptance model as an organizing framework. *Online Learning*, 21(1), 15–35.

Winters, J.V. (2011). Why are smart cities growing? Who moves and who stays. *Journal of Regional Science*, 51(2), 253–270.

Zhao, Y. and Watterston, J. (2021). The changes we need: Education post COVID-19. *Journal of Educational Change*, 22(1), 3–12.

Conclusion

Shaping the Digitized City

Not only may every activity be comprehended and defined to the extent that it assumes form and inscribes its graph in space and time, but life itself, furthermore, is essentially a creator of forms. Life is form, and form is the modality of life.

Henri Focillon, 1943

C.1. Introduction

The failure of the "smart city" slogan, at least in the world's democracies, has not canceled, slowed down, or transformed the procedural logic of the digitization of city life (Merveille 2022).

To appreciate this reality, we will invoke the illustrious Alain Supiot who, in his lecture at the French Society of Architects in 2021, distinguished the terms "city" and "town". He explained, during his presentation, that "the city does not designate an agglomeration, it is not a concentration of population in a given territory, but the association of citizens under the aegis of a common law" (Supiot 2021)[1].

In this sense, he continues, living together is not enough to form a society. Here again, Alain Supiot invites us to distinguish between the notions of living together and forming a society. Living together refers to the idea of adding up the individuals

Conclusion written by Nicolas MERVEILLE.

1 Lecture by Alain Supiot, *Le crédit de la parole*. French Society of Architects, Thursday, December 2, 2021. See: https://www.youtube.com/watch?v=KgCbWkgVI1k.

who occupy a given physical space. Thus, we live together, often indifferent to each other, in the same neighborhood or the same countryside.

On the other hand, forming a society, says the master, is a matter of intentionality. It is the intention of citizens to bind their fate together under the aegis of a common law. In other words, forming a society means taking root. It is situated in identity. To conclude this idea, we will specify that we form a society in the city.

Throughout the text, the working hypothesis we will explore is that the digitized city is now the social project of a globalized mass, where the fate of individuals essentially rests on the procedural logic of digitization.

What is digitization? We adopt Lucas' (2014) definition of digitization: "The operation of digitization consists of translating information into numbers, based on a binary coding made up of 0s and 1s" (p. 80). Digitization is a term covering a plurality of technical instruments and computer code that enable duplication of a physical object or information to produce its digital representation.

What does digitization do to the city? This is the problem that we have explored slowly and scrupulously over several years (Caccamo et al. 2019). We are aware that we are only at the beginning of systematic digitization of the lives of city dwellers, and we adhere to the argument that any attempt to describe the form of the digitized city would be doomed to failure. On the other hand, it seems reasonable to us to follow in the footsteps of Donna Haraway, Eric Hughes, J.G. Ballard, Masamune Shirow, Flore Vasseur and Rob Kitchin, who have already explored it at length and for whom, contrary to what is endlessly repeated to contemporary citizens, the current upheaval in the order of things is of an anthropological rather than a technoscientific nature (Gori 2020).

Due to the rhythms and nature of contemporary technical innovations, it is the attributes we have given ourselves as humans that are called upon to change (Virilio 2009). However, the smart era is not sufficiently equipped to explain the meaning of what is happening. We are all led to experience the reality of Jacques Tati in his film *Mon Oncle*. Our ideas of the world are continuously challenged, and the invitation to change has quickly become a form of technocratic authoritarianism (Stiegler 2019). In other words, we are experiencing an acceleration of technoscientific innovation, which is playing a major role in transforming the idea that we collectively have of ourselves.

Our choice of format to structure our argument is the essay. In this, you will find no trace of a description of a proven methodology or the beginning of a quantified demonstration. We share Gori's (2020) and Sadin's (2011) opinion that new universes have arisen and that we are required to create new paradigms to evaluate them.

Here are the five notions that we will discuss in the next few pages, which we consider to be the organizing schemas that convert the analog city into a digitized city: form, territory, value, formula and tokenization. We do not claim that the advent of the digital city is limited to these five notions. This is just a first glimpse of an effort that will take several decades.

C.2. Formation

Form is the unity of multiplicity. It is defined by Wolff (2010) as "the arrangement of material elements which gives the whole its structural unity and allows it to exercise its proper function, its purpose" (p. 45). More recently, Tim Ingold reaffirms that:

> Form is an emergent property of a global system constituted by the relationships between diverse components. A change in any component of the system, whether the genome or some aspect of the intra- or extra-organismic environment, to the extent that it alters the parameters of development, may lead to significant changes in form. The possibilities for change are not unlimited, however; they are limited by the range of forms that can be generated by the properties of the dynamic system of organization (Ingold 2013, p. 67).

By evoking form to specify the meaning of the current solidarities between the whole and the parts, Adorno articulates science, the arts and philosophy with management. It is therefore a question of discovering and comparing the invariant element of a system of transformation (D'Arcy Thompson 1994). Inertias are induced by a plurality of conjunctures. Francis Wolff again insists on the possibility of using form to understand complex dynamics:

> It is through form thus understood that we can rationally apprehend substances, name them, classify them, describe them, explain them, in summary, make science of them (Wolff 2010, p. 52).

For Maurice Merleau-Ponty, on page 143 of his masterpiece *The Structure of Behavior*, form is constitutive of three categories: quantity, order and value (or meaning). If we follow Merleau-Ponty, we note that conceiving of a form implies three operations: quantifying relationships, describing hierarchies and qualifying what is being established in it and for it.

The configuration of current solidarities between the parties will generate at time *t* a form whose meaning may be evaluated and subsequently compared to other types of forms. Examining a form and the evolution of its transformations during its life cycle should, in the long term, facilitate decision-making processes at various points in time.

> The life of forms is not the result of chance. Nor is it a great cyclorama neatly fitted into the theater of history and called into being by historical necessities. No. Forms obey their own rules – rules that are inherent in the forms themselves – and there is no reason why we should not undertake an investigation of how these great ensembles, united by close reasoning and by coherent experiment, behave through the phases that we call their life (Focillon 1943, p. 21).

We will clarify immediately that in evoking form, we are referring to Paul Klee's intuition, which we discovered in Ingold (2013), about formation: "Form is death" while "formation is life" (Klee 1977)!

As Ingold (2013, p. 221) writes, although both perspectives are imbued with holism, they fundamentally diverge in their appreciation of the whole. Form evokes a completed (closed) whole, while formation evokes the (unfinished) process. It is worth mentioning, since we will use this argument in the following lines, that for Umberto Eco, "whoever says form says accomplishment". Form must be used as a completed principle (Sorlin 1998).

We already read this idea in the writing of George Simmel who differentiates the terms *form* and *formung* (the action of giving a form). It was further developed by Henri Maldiney who opposed form as made (*Gestalt*) to form being made (*Gestaltung*). We found this magnified in Bailly (2013, p. 32), for whom: "every form, and this is fundamental, is only a state of form". It is therefore within this second tradition, that of a form which is being formed, that we locate ourselves, because it emphasizes a perpetual dynamic at work. For us, it is therefore only possible to describe the form of the analog city, which has already been completed, and of which it is now possible to embrace the whole, and to detect in its shadow the first indications of the formation of the digital city.

William James, inspired by Henri Bergson, wrote: "What really exists is not things made, but things in the making. Once made they are dead, and an infinite number of alternative conceptual decompositions can be used in defining them" (James 2007, p. 177).

Now that the intention of this paper has been formalized, we embark on the explanation of these five things that are in the process of forming: form, territory, value, formula and tokenization.

C.3. The interweaving of territories

For Immanuel Kant, life is the first purpose of living creatures. The organic therefore is the bearer of a project that can only be realized within its territory. A living creature is incapable of living in space, and this is why it produces territory. Production of territory happens through the modalities of appropriation of space. A territory is the result of a modality of appropriation of space, which is specific to each species.

Species are sometimes in competition, sometimes in co-option and often in cooperation in producing their territories. As for *Homo sapiens sapiens*, and despite our deep respect for geographers and archaeologists, we will refer, once again, to the masterpiece by the anthropologist Maurice Godelier entitled *L'ideel et le materiel* (*The Ideal and the Material*).

Maurice Godelier explains that *Homo sapiens sapiens* appropriates space through three dimensions: the subjective dimension (culture), the concrete dimension (use) and the abstract dimension (legal). It is through the realization of these three dimensions that *Homo sapiens sapiens* creates its territory, where it becomes possible for it to realize its life project.

Appropriation of space is dynamic and permanent. It is as much conflictual as it is consensual. We go to war, then we reconcile. In the digitized city, it is no longer simply a question of conventional warfare, but also of cybernetic warfare.

The issues of organizational security are increasingly digital. There is little doubt that geopolitical issues have become more complex due to the institutionalization of digitization.

Yet, too often, we still read that the imaginary, the virtual and the material are parallel spaces. They are presented to us as three hermetic and autonomous dimensions, where each one's logic of functioning and (de)regulation is completely foreign to the others. This is also the suspicion that governs the mechanisms of reasoning that aim to clarify the nature of what they comprise. For makers of public opinion, the *dark web* is necessarily a mafia operation, cryptocurrency finances terrorism and the disappearance of cash is inevitable.

Yet, when we move from one space to another, we experience a universe familiar to us, since we find there the cosmological invariants that have shaped the collective imagination of consumerism.

It allow us to advance the hypothesis that the formation of the digital city probably began with the interlocking of the territories of the imaginary, the virtual and the material, which now function in total interdependence and synchronicity through the blockchain.

It was in the year 2009 that the first operational blockchain saw the light of day. It was named *bitcoin*. Its creator(s) have remained anonymous. A posture that reflects the creation: refined.

Virtual territory has never been a parallel universe, since it follows the modalities of appropriation of physical space. Physical space is not disconnected from the imaginary, since it is the product of a subjective dimension, which is itself constitutive of the imaginary.

What is happening is the emergence of a digitized identity that is beginning to deform the analog city, since the values[2] generated in one of these three territories can freely migrate to the other two.

Elon Musk has tried to simultaneously appropriate all three of these territories to establish his legitimacy over the next stages of the formation of the digital city: the imaginary (Neuralink), the virtual (Twitter) and the material (SpaceX).

To complete his project, Elon Musk acquired the cryptocurrency *Dogecoin* in 2022, whose value he can now increase or decrease with a simple tweet.

Even before acquiring it contractually, in the same year, Elon Musk set about appropriating the Twitter network by redefining the three modalities of appropriation of space through tweeting: freedom of expression (subjective dimension), cancellation of moderation mechanisms (concrete dimension) and the operationalization of network deregulation outside of any existing legal framework (abstract dimension).

C.4. Reconfiguration of the circulation of value

Creation of value in the territories of the imaginary and the virtual is not specific to the digital city. In fact, this phenomenon has been described in the analog city due to the popularity of the video game *Second Life*[3].

2 Economic, heritage, social and religious.

3 See: https://secondlife.com.

On the other hand, the free circulation of value within these three interdependent and synchronized territories is a novel feature of the digitized city. Like Elon Musk's consortium, we must appreciate the metamorphoses of the economic, the social and the political in their four dimensions: imaginary, virtual, material and temporal.

Since the creation of the bitcoin blockchain, the bullish and bearish phases of the value of the bitcoin token have been spread over a four-year cycle. The argument put forward to explain this regularity is the consequence of the Halving[4] mechanism, meaning that bitcoin miners see their bonuses halved every four years, leading to a systematization of scarcity in the quantity of bitcoins available.

The dates of the *Halving* are easy to determine, and the interests involved organize themselves accordingly, with better or worse luck. When we examine the trajectory of bitcoin millionaires, we see that they do not have many options to reinvest their earnings in the so-called real economy because they do not generally benefit from the social status necessary to access more profitable markets. Lacking political power, but with financial capital in the bank, they naturally adopt stereotypical purchasing behaviors modeled on the lifestyles of the *nouveau riche*: big cars, haute couture goods and villas with swimming pools.

However, traditional analyses of the state of the real estate and automobile markets generally ignore this client base. It is hastily described as marginal by experts who have no understanding of the modalities of circulation of value in the digital city. And yet, real estate plays an important role in bitcoin's speculative cycle. Why? The best traders manage to buy bitcoin when its financial value has dropped by nearly 80% from its last high point.

Some of them at that time invest the entire capital acquired from selling their home, which they themselves call a down payment. Less than 24 months later, they will enjoy a comfortable capital gain. The logic behind the operation of bitcoin has nothing to do with the lottery or the casino. It is less risky, because it is less uncertain! Even before the bearish phase has started, they sell their bitcoin tokens to buy a better, and necessarily more expensive, home, in which they can live in peace through the next nuclear winter of bitcoin. Buying and selling bitcoin tokens very often speeds up the rate of capital accumulation and circulation. The dynamism of the real estate market in Montreal can be explained by multiple demographic, socioeconomic and geopolitical factors that are often understood by real estate

4 See: https://www.investopedia.com/bitcoin-halving-4843769#toc-when-did-the-bitcoin-halvings-happen.

brokers. On the other hand, they are still unaware of the speculative cycle of bitcoin that influences the evolution of prices in a market, even though they thought they understood all of its workings. The great media experts of the analog city are almost always incompetent when it comes to the digital city.

In the analog city, teenagers bought game consoles and paid the gaming industry for video games. This business model was structured on the logic of pay to play. Now, in the play to earn business model, players expect to receive compensation in cryptocurrency after solving a puzzle, advancing to a higher level, or completing a game. If they play correctly, they recoup their initial investment and then resell their console to a lost soul still stuck in the logic of pay to play.

The best players enjoy star status within gamer communities and generate enough income to buy housing with their bitcoin tokens. Their games are streamed live on a Twitch platform channel, and in return, they receive revenue for advertising contracts. When we talk to their parents, we understand that they regret the time their offspring waste on this damned console. They are far from imagining that their teenagers are building up a small nest egg via the logic of Web 3.0.

Up until now, we have been talking about economic and financial value, but it is the whole cultural framework that is being transformed, and values are being created, diffused and shared through our artifacts and algorithmic governance. The great institutions of the analog city (family, state, school, company and association) no longer have a monopoly on the manufacture, selection and dissemination of values. In the digitized city, it is the algorithm that has been enthroned to govern human affairs (Rouvroy and Bernes 2013).

C.5. The performative formula

An algorithm is a formula. It was while reading Alice Krieg-Planque's work on the operations of the formula in the discursive context that the idea came to us of putting into perspective the *modus operandi* of the algorithm over the city. The criteria chosen by Alice Krieg-Planque to capture what the formula does to discourse reminded us of the writings of Antoinette Rouvroy concerning what the algorithm does to the analog city. This also seemed to apply to Rob Kitchin's seminal work on the algorithm and the metamorphoses of the urban.

For both these monuments of thought, algorithms structure behaviors, habits and lifestyles. So, what is a formula?

Alice Krieg-Planque defined it in terms generic enough for us to evoke in turn:

> A set of formulations which, due to their use at a given time and in a given public space, crystallize political and social issues which these expressions at the same time contribute to constructing (Krieg-Planque 2009, p. 7).

Mayaffre (2009), in his commentary on Alice Krieg-Planque's work, notes that in Chapter 4 of her book a relevant characterization of the notion of formula is proposed: stability of the signifier, discursive dimension, status of social referent and polemical aspect.

Stability of the signifier is the sine qua non condition for the formula to be taken up and circulated. In this regard, Alice Krieg-Planque takes up the term "freezing" from linguists. The discursive dimension refers to the impossibility of appreciating the meaning of the formula outside of the statement-situation dyad:

> In other words, it is social usage that constructs and institutes it as a formula, and not some predictable linguistic codes or rules. In this, a formula is treated as a discourse, in other words, by thinking about the reflexive tension between the text and the socio-historical context, the discourse and the political facts (Mayaffre 2009, p. 2).

The status of social referent emanates from the normative and contagious characteristics (Sperber 1996) of the formula, which may be converted into despotism:

> Thus, today we could not avoid the formulas of sustainable development or renewable energy in the debate on ecology. Even if a speaker were to reject these formulas, they would still be situated in relation to them; even if they denounced them, they would participate in their circulation (Mayaffre 2009, p. 2).

The formula is a source of polemics. This is explained by its disjunctive nature. We know it to be predatory (the formula of black magic), but we also know it to be emancipatory (the mathematical formula).

> There has been endless discussion about this little text, wondering whether Bartleby was mad, cynical, inconsistent, or rebellious. Of all of these interpretations, we will neither retain nor reject any. For Melville's story does nothing but say what it says, literally and in

> every sense. And what it says is that magic formulas exist. That some magic formulas have the power to change the order of things, while others, on the contrary, are formulas complicit with the powers that be in reinforcing this order of things (England and D'Elia)[5].

The keystone of Alice Krieg-Planque's work, which we retranslate into our own words, is that the formula works. In this regard, Gilles Deleuze wrote that: "the formula is valid as a process"[6]. Thus, we anchor this reflection in the theory of speech acts that arises from the work of John Austin. Following Austin (1991), it goes without saying that the formula is performative, in other words, it acts as:

> A performative statement, by the mere fact of its utterance, accomplishes the action concerned: it is enough for the chair of a meeting to say 'I declare the meeting open' to effectively open the meeting. The performative statement is therefore opposed to the constative statement which simply describes an action whose execution is, moreover, independent of the utterance: saying "I open the window" does not, ipso facto, open the window, but describes an action. The performative statement is therefore both a linguistic manifestation and an act in reality (Catherine Fuchs, "Actes de langage", *Encyclopédie Universalis*).

The promoters of algorithms experiment with the emancipatory register of the action of the formula, while the scrupulous explore its predatory register. Collectives such as the Quadrature du Net have been formed to raise public awareness that the algorithms that preside over the formation of the digitized city are silent about what they do and do not do what they say.

Unfortunately, the antagonistic nature of the formula is hardly understood by elected politicians, and reflections continue to be polarized between technophiles and technophobes. Affirming that the algorithm is not neutral is a step in the right direction, but it is not enough. Every formula carries an intention, which ipso facto neutralizes any idea of neutrality.

However, recognizing its antagonistic nature would create the conditions for a public debate around constructing new mechanisms of regulation and control over their actions.

---

5 January 27, 2008, "Pouvoirs des formules magiques/formules magiques des pouvoirs", see: endehors.net.

6 Deleuze, G. (1989). Afterword. In *Bartleby*, Melville, H. (ed.). Flammarion, Paris.

The Biosphere Code Manifesto[7] initiative, which was launched in October 2015 in Stockholm, addresses the issue of the status and roles of algorithms over the biosphere from the perspective of social responsibility. It seeks to build a political platform where the future of the biosphere of the digitized city will be debated, as it understands how performative the formula is and that its modus operandi takes effect over the three entangled territories: virtual, material and imaginary, affecting both digital operating systems and all living species on the globe.

C.6. Tokenization of the biosphere and cryptoconservation

Tokenization is a computer programming procedure that aims to create a digital representation of an asset, that is, a token, which will be mined, stored and traded on a blockchain.

The term "asset" encompasses anything that has monetary value to at least two parties and may be exchanged with another asset of the same type, meaning it is fungible. An asset may be a biological organ, an ecological or mechanical service, a physical object or a right.

The smart contract is formulated in computer code that enables its autonomous execution as soon as certain conditions are met, as well as its immutability. In other words, the codification of a contract in blockchain technology offers an opportunity to execute the terms of a contract from the virtual territory into the territory of materiality.

The range of applications is already very broad (payment, real estate, inheritance and insurance). Yesterday, in the past, it was possible to acquire movable property by paying for it with bitcoin tokens.

Now, it is real estate that is tokenized, bought and sold on the new exchange infrastructure: the blockchain.

We readily acknowledge that it is counterintuitive to argue that the urban environment has never been the driver of disruption. And yet, the countryside has been a playground for the multinationals of the digital city for longer than the urban areas. John Deere tractors and combine harvesters[8] are autonomous and have been plundering the knowledge of peasants for a long time. Precision agriculture was

7 See: https://nextnature.net/story/2015/the-biosphere-code-manifesto#:~:text=The%20precepts%20for%20an%20in,ecosystems%22%20as%20The%20Guardian%20wrote.

8 See: https://www.vice.com/en/article/xykkkd/why-american-farmers-are-hacking-their-tractors-with-ukrainian-firmware.

already using drones in the early 2000s to assess the quality of soil on farms covering several hundred hectares per cultivable plot. Artificial intelligence is everywhere in animal production and the agrofuel sector.

Legal, sociological, political and ethical constraints over technological deployment are less onerous in regions with low population density. All of the ecosystems of the globe are now smart, since a digital infrastructure has been grafted onto them via the Internet of Things. The high seas[9] are a future theater of operations for transhumanism. In short, the smart city is nothing more than a rebellious and suicidal offshoot of the digital city.

Tech4All is an initiative from Huawei, a Chinese company specializing in information and communication technologies. According to Huawei, Tech4All aims to help developing countries gain access to information and communications technologies to help them develop their economies and improve the well-being of their populations.

In 2020, the IUCN Global Program on Protected and Conserved Areas signed an agreement with Huawei, called TECH4NATURE[10]. At the heart of the partnership is the IUCN's Green List of Protected and Conserved Areas standard. Since then, we have seen the emergence of a multitude of initiatives based on the tokenization of biodiversity and iconic species.

By 2023, Tech4Nature has set a goal of enabling more than 300 protected areas in more than 30 countries worldwide to assess their conservation success using the IUCN Green List standard. The World Wide Fund for Nature (WWF)[11], Conservation International (CI)[12] and the Big International Non-Governmental Organizations (BINGOs) of conservation promote the blockchain and the tokenization of living things to try to protect biodiversity. The blockchain is seen by ecologists in these organizations as the new infrastructure for capturing and allocating capital to support their biosphere conservation programs. We are far from the imaginary of the radicals of political ecology.

9 French Polynesia and the Seasteading Institute (see: https://www.seasteading.org) have provided for the establishment of an island and the development of a special legal regime. To explore this type of initiative in more depth, read the following excellent report: Cayol. A. and Gaillard, E. (eds) (2002). Trans-humanisme(s) & droit(s). Report, Institut des études et de la recherche sur le droit et la justice, p. 283.

10 See: https://panorama.solutions/sites/default/files/mediathek/panorama_tech4nature_-_v09_final_web_002.pdf.

11 See: https://techhub.wwf.ca/technology/blockchain-applications/.

12 See: https://medium.com/gochain/conservation-international-joins-gochain-network-as-a-blockchain-signing-node-gochain-b416dc9fd6af.

C.7. Conclusion

The era has given birth to a damning metaphor, found in the mouths of the promoters of the smart city, the slogan urging citizens to "think outside the box". This is a level of perversion never achieved before, because what is sought is that we turn our gaze away from what is happening in the formation of the digital city.

When we look inside this box, we are struck by what we find there, and we see that the new elements that have found a home there have ended up destroying the original structure of the analog city. In describing the box, thinking about the box and analyzing its contents should be our priority to better describe the nature of the issues that our city is facing.

C.8. References

Adorno, T.W. (1995). *Théorie esthétique*. Klincksieck, Paris.

Austin, L.J. (1991). *Quand dire, c'est faire*. Le Seuil, Paris.

Bailly, J.C. (2013). *Sur la forme*. Manuella, Paris.

Caccamo, E., Walzberg, J., Reigeluth, T., Merveille, N. (2019). *De la ville intelligente à la ville intelligible*. Presses de l'Université du Québec, Montreal.

Deleuze, G. (1989). Postface. In *Bartleby*, Melville, H. (ed.). Flammarion, Paris.

Eco, U. (1965). *L'oeuvre ouverte*. Le Seuil, Paris.

Focillon, H. (1981). *La vie des formes*. Presses Universitaires de France, Paris.

Godelier, M. (1984). *L'idéel et le matériel*. Éditeur Fayard, Paris.

Gori, R. (2020). *Et si l'effondrement avait déjà eu lieu – l'étrange défaite de nos croyances*. Éditions Liens Qui Liberent, Paris.

Ingold, T. (2013). *Marcher avec les dragons*. Éditions Zones Sensibles, Paris.

James, W. (2007). *Philosophie de l'expérience. Un univers pluraliste*. Gallimard, Paris.

Klee, P. (1977). *Écrits sur l'art*. Dessain et Tolra, Paris.

Krieg-Planque, A. (2009). *La notion de "formule" en analyse du discours. Cadre théorique et méthodologique*. Presses Universitaires de Franche-Comté, Besançon.

Lucas, J.-F. (2014). La numérisation de la ville et ses représentations. *Magazine des Cultures Digitales*, 73, 78–85.

Mayaffre, D. (2009). *Commentaire sur Alice Krieg-Planque. La notion de "formule" en analyse du discours. Cadre théorique et méthodologique.* Presses Universitaires de Franche-Comté, Besançon.

Merveille, N. (2002). Une réflexion anthropologique sur la cité numérisée. *VertigO la revue électronique en sciences de l'environnement*, Special Issue(36). doi: 10.4000/vertigo. 37323.

Rouvroy, A. and Berns, T. (2013). Gouvernementalité algorithmique et perspectives d'émancipation. Le disparate comme condition d'individuation par la relation ? *Réseaux*, 177(1), 163–196.

Sadin, E. (2011). *Société de l'anticipation : le web précognitif ou la rupture anthropologique.* Inculte, Paris.

Sorlin, P. (1998). *La Forme en jeu.* Presse Universitaires de Vincennes, Vincennes.

Stiegler, B. (2019). *Il faut s'adapter : sur un nouvel impératif politique.* Éditions Gallimard, Paris.

Thompson, D.A. (1994). *Forme et Croissance.* Le Seuil, Paris.

Virilio, P. (2009). *Le futurisme de l'instant.* Éditions Galilée, Paris.

Wolff, F. (2010). *Notre humanité : d'Aristote aux neurosciences.* Fayard, Paris.

List of Authors

Intissar ABBES
PRESTIGE
IHEC
Carthage
Tunisia

Isabelle BORIES-AZEAU
MRM
Université de Montpellier
France

Léa CAUCHARD
BRP
Montreal
Canada

Blandine CHAPEL
MRM
Université de Montpellier
France

Jessica CLEMENT
CIRAD
Montpellier
France

Philippe COHARD
MRM
Université de Montpellier
France

Fabien DURIF
OCR GreenUXlab
UQAM
Montreal
Canada

Pauline FOLCHER
MRM
Université de Montpellier
France

Ambre GAMBIN
MRM
Université de Montpellier
France

Karine GARCIA
MRM
Université de Montpellier
France

Sarah GARIDI
CEREFIGE
Université de Lorraine
Metz
France

Olga GONCALVES
UPVD
Perpignan
France

Andréa GOURMELEN
MRM
Université de Montpellier
France

Julie JAMMES
MRM
Université de Montpellier
France

Inès KOLLI
UQAM
Montreal
Canada

Camille LACAN
CRESEM
UPVD
Perpignan
France

Sara LAURENT
Montpellier Business School
France

Arnaud MAES
MRM
Université de Montpellier
France

Nicolas MERVEILLE
CIRODD
ISE
Université du Québec
Montreal
Canada

Sarah MUSSOL
MRM
Université de Montpellier
France

Gilles N'GOALA
MRM
Université de Montpellier
France

Florence PAULHIAC SCHERRER
Chaire internationale sur les usages et pratiques de la ville intelligente
UQAM
Montreal
Canada

Audrey PORTES
Montpellier Business School
France

Angélique RODHAIN
MRM
Université de Montpellier
France

Index

A, C

D, E

G, I

L, M

P, S

T, U

Other titles from

iSTE

in

Science, Society and New Technologies

2025

CHRISTOFLE Sylvie
Tourism Agility in Times of Crisis and Uncertainty 1: Between Technology and Society
(Tourism and Mobility Systems Set – Volume 11)

CHRISTOFLE Sylvie
Tourism Agility in Times of Crisis and Uncertainty 2: Between Marketing-Communication and Sustainability
(Tourism and Mobility Systems Set – Volume 12)

LEVEAU Pierre-Henry
Tourism, Technologies and Consumption in the 5.0 Era: Discovery and Reflection in the Face of Organizational and Societal Issues
(Tourism and Mobility Systems Set – Volume 10)

2024

BARTHES Angela, CIBIEN Catherine, ROMAGNY Bruno
Biosphere Reserves and Sustainable Development Goals 1: Scientific and Practical Educational Issues in the Mediterranean
(Territory Development Set – Volume 3)
Biosphere Reserves and Sustainable Development Goals 2: Issues, Tensions, Processes and Governance in the Mediterranean
(Territory Development Set – Volume 4)

BOTERO Nataly, LEDOUBLE Hélène, ALLARD-HUVER François
Pesticides: Pluridisciplinary Dialogues in Social and Human Sciences
(Communication, Environment, Science and Society Set – Volume 4)

CATELLANI Andrea, LITS Grégoire
Transitions in Tension: Controversies and Tensions Around Ecological Transitions
(Communication, Environment, Science and Society Set – Volume 3)

DAVALLON Jean
Heritage Traces in the Making: A Communicational Analysis of Modes of Heritagization
(Traces Set – Volume 6)

DEJEAN Maxime, L'HOSTIS Marine, LI Meng
Chinese Outbound Tourism: From Source Markets to Destinations
(Tourism and Mobility Systems Set – Volume 8)

GAY Jean-Christophe
Tourismophobia: From "Mass Tourism" to "Overtourism"
(Tourism and Mobility Systems Set – Volume 9)

JULIA Guilhem, FAUCHON Anne, KANAWATI Rushed
AI, Healthcare and Law

KOHLMANN Émilie
Linking with Nature in the Digital Age
(Communication, Environment, Science and Society Set – Volume 1)

LEDOUBLE Hélène
Popularizing Science: The Complex Terminological Interactions between Scientific and Press Discourses within the Field of Agroecology
(Communication, Environment, Science and Society Set – Volume 2)

LOUART Pierre, PIRIOU Jérôme, VIOLIER Philippe
Tourist Destinations According to Stakeholder Strategies: From Clusters to Local Tourist Systems
(Tourism and Mobility Systems Set – Volume 7)

2023

APCHAIN Thomas
Favela Tours: Building Otherness in the Favelas of Rio de Janeiro
(Tourism and Mobility Systems Set – Volume 6)

CARDON Alain, EL HAMI Abdelkhalak
Fundamental Generation Systems: Computer Science and Artificial Consciousness, the Informational Field of Generation of the Universe, the Sixth Sense of Living Beings
(Digital Sciences Set – Volume 4)

CURULLI Irene G., IKIZ KAYA Deniz, KHAEFI Arghavan
Heritage Education for Climate Action
(Research in Architectural Education Set – Volume 2)

ELAMÉ Esoh
The Sustainable City in Africa Facing the Challenge of Liquid Sanitation
(Territory Development Set – Volume 2)

JURCZENKO Emmanuel
Climate Investing: New Strategies and Implementation Challenges

LYUBAREVA Inna, WALDECK Roger
Diversity of Methodological Approaches in Social Sciences: Example of the Analysis of Media and Online Information
(Modeling Methodologies in Social Sciences Set – Volume 3)

PARET Dominique, CRÉGO Pierre, SOLÈRE Pauline
Smart Patches: Biosensors, Graphene and Intra-Body Communications

VENTRE Daniel, GUILLOT Philippe
Electronic Communication Interception Technologies and Issues of Power

VENTRE Daniel, LOISEAU Hugo
Cybercrime during the SARS-CoV-2 Pandemic (2019-2022): Evolutions, Adaptations, Consequences
(Cybersecurity Set – Volume 3)

2022

AIT HADDOU Hassan, TOUBANOS Dimitri, VILLIEN Philippe
Ecological Transition in Education and Research

CARDON Alain
Information Organization of The Universe and Living Things: Generation of Space, Quantum and Molecular Elements, Coactive Generation of Living Organisms and Multiagent Model
(Digital Sciences Set – Volume 3)

CAULI Marie, FAVIER Laurence, JEANNAS Jean-Yves
Digital Dictionary

DAVERNE-BAILLY Carole, WITTORSKI Richard
Research Methodology in Education and Training: Postures, Practices and Forms
(Education Set – Volume 12)

ELAMÉ Esoh
Sustainable Intercultural Urbanism at the Service of the African City of Tomorrow
(Territory Development Set – Volume 1)

FLEURET Sébastien
A Back and Forth Between Tourism and Health: From Medical Tourism to Global Health
(Tourism and Mobility Systems Set – Volume 5)

KAMPELIS Nikos, KOLOKOTSA Denia
Smart Zero-energy Buildings and Communities for Smart Grids
(Engineering, Energy and Architecture Set – Volume 9)

2021

BARDIOT Clarisse
Performing Arts and Digital Humanities: From Traces to Data
(Traces Set – Volume 5)

BENSRHAIR Abdelaziz, BAPIN Thierry
From AI to Autonomous and Connected Vehicles: Advanced Driver-Assistance Systems (ADAS)
(Digital Sciences Set – Volume 2)

BÉRANGER Jérôme
Societal Responsibility of Artificial Intelligence: Towards an Ethical and Eco-responsible AI
(Technological Prospects and Social Applications Set – Volume 4)

CORDELIER Benoit, GALIBERT Oliver
Digital Health Communications
(Technological Prospective and Social Applications Set – Volume 5)

DOUAY Nicolas, MINJA Michael
Urban Planning for Transitions

GALINON-MÉLÉNEC Béatrice
The Trace Odyssey 1: A Journey Beyond Appearances
(Traces Set – Volume 4)

HENRY Antoine
Platform and Collective Intelligence: Digital Ecosystem of Organizations

KOLOKOTSA Denia, KAMPELIS Nikos
Smart Buildings, Smart Communities and Demand Response
(Engineering, Energy and Architecture Set – Volume 8)

LE LAY Stéphane, SAVIGNAC Emmanuelle, LÉNEL Pierre, FRANCES Jean
The Gamification of Society
(Research, Innovative Theories and Methods in SSH Set – Volume 2)

RADI Bouchaïb, EL HAMI Abdelkhalak
Optimizations and Programming: Linear, Non-linear, Dynamic, Stochastic and Applications with Matlab
(Digital Sciences Set – Volume 1)

2020

ALAKTIF Jamila, CALLENS Stéphane
Migration and Climate Change: From the Emergence of Human Cultures to Contemporary Management in Organizations

BARNOUIN Jacques
The World's Construction Mechanism: Trajectories, Imbalances and the Future of Societies
(Interdisciplinarity Between Biological Sciences and Social Sciences Set – Volume 4)

ÇAĞLAR Nur, CURULLI Irene G., SIPAHIOĞLU Işıl Ruhi, MAVROMATIDIS Lazaros E.
Thresholds in Architectural Education
(Engineering, Energy and Architecture Set – Volume 7)

DUBOIS Michel J.F.
Humans in the Making: In the Beginning was Technique
(Social Interdisciplinarity Set – Volume 4)

ETCHEVERRIA Olivier
The Restaurant, A Geographical Approach: From Invention to Gourmet Tourist Destinations
(Tourism and Mobility Systems Set – Volume 3)

GREFE Gwenaëlle, PEYRAT-GUILLARD Dominique
Shapes of Tourism Employment: HRM in the Worlds of Hotels and Air Transport
(Tourism and Mobility Systems Set – Volume 4)

JEANNERET Yves
The Trace Factory
(Traces Set – Volume 3)

KATSAFADOS Petros, MAVROMATIDIS Elias, SPYROU Christos
Numerical Weather Prediction and Data Assimilation
(Engineering, Energy and Architecture Set – Volume 6)

MARTI Caroline
Cultural Mediations of Brands: Unadvertization and Quest for Authority
(Communication Approaches to Commercial Mediation Set – Volume 1)

MAVROMATIDIS Lazaros
Climatic Heterotopias as Spaces of Inclusion: Sew Up the Urban Fabric
(Research in Architectural Education Set – Volume 1)

MOURATIDOU Eleni
Re-presentation Policies of the Fashion Industry: Discourse, Apparatus and Power
(Communication Approaches to Commercial Mediation Set – Volume 2)

SCHMITT Daniel, THÉBAULT Marine, BURCZYKOWSKI Ludovic
Image Beyond the Screen: Projection Mapping

VIOLIER Philippe, with the collaboration of TAUNAY Benjamin
The Tourist Places of the World
(Tourism and Mobility Systems Set – Volume 2)

2019

BRIANÇON Muriel
The Meaning of Otherness in Education: Stakes, Forms, Process, Thoughts and Transfers
(Education Set – Volume 3)

DESCHAMPS Jacqueline
Mediation: A Concept for Information and Communication Sciences
(Concepts to Conceive 21st Century Society Set – Volume 1)

DOUSSET Laurent, PARK Sejin, GUILLE-ESCURET Georges
Kinship, Ecology and History: Renewal of Conjectures
(Interdisciplinarity Between Biological Sciences and Social Sciences Set – Volume 3)

DUPONT Olivier
Power: A Concept for Information and Communication Sciences
(Concepts to Conceive 21st Century Society Set – Volume 2)

FERRARATO Coline
Prospective Philosophy of Software: A Simondonian Study
(Technological Prospects and Social Applications Set – Volume 3)

GUAAYBESS Tourya
The Media in Arab Countries: From Development Theories to Cooperation Policies

HAGÈGE Hélène
Education for Responsibility
(Education Set – Volume 4)

LARDELLIER Pascal
The Ritual Institution of Society
(Traces Set – Volume 2)

LARROCHE Valérie
The Dispositif: A Concept for Information and Communication Sciences
(Concepts to Conceive 21st Century Society Set – Volume 3)

LATERRASSE Jean
Transport and Town Planning: The City in Search of Sustainable Development

LELEU-MERVIEL Sylvie, SCHMITT Daniel, USEILLE Philippe
From UXD to LivXD: Living eXperience Design

LENOIR Virgil Cristian
Ethically Structured Processes
(Innovation and Responsibility Set – Volume 4)

LOPEZ Fanny, PELLEGRINO Margot, COUTARD Olivier
Local Energy Autonomy: Spaces, Scales, Politics
(Urban Engineering Set – Volume 1)

METZGER Jean-Paul
Discourse: A Concept for Information and Communication Sciences
(Concepts to Conceive 21st Century Society Set – Volume 4)

MICHA Irini, VAIOU Dina
Alternative Takes to the City
(Engineering, Energy and Architecture Set – Volume 5)

PÉLISSIER Chrysta
Learner Support in Online Learning Environments

PIETTE Albert
Theoretical Anthropology or How to Observe a Human Being
(Research, Innovative Theories and Methods in SSH Set – Volume 1)

PIRIOU Jérôme
The Tourist Region: A Co-Construction of Tourism Stakeholders
(Tourism and Mobility Systems Set – Volume 1)

PUMAIN Denise
Geographical Modeling: Cities and Territories
(Modeling Methodologies in Social Sciences Set – Volume 2)

WALDECK Roger
Methods and Interdisciplinarity
(Modeling Methodologies in Social Sciences Set – Volume 1)

2018

BARTHES Angela, CHAMPOLLION Pierre, ALPE Yves
Evolutions of the Complex Relationship Between Education and Territories
(Education Set – Volume 1)

BÉRANGER Jérôme
The Algorithmic Code of Ethics: Ethics at the Bedside of the Digital Revolution
(Technological Prospects and Social Applications Set – Volume 2)

DUGUÉ Bernard
Time, Emergences and Communications
(Engineering, Energy and Architecture Set – Volume 4)

GEORGANTOPOULOU Christina G., GEORGANTOPOULOS George A.
Fluid Mechanics in Channel, Pipe and Aerodynamic Design Geometries 1
(Engineering, Energy and Architecture Set – Volume 2)
Fluid Mechanics in Channel, Pipe and Aerodynamic Design Geometries 2
(Engineering, Energy and Architecture Set – Volume 3)

GUILLE-ESCURET Georges
Social Structures and Natural Systems: Is a Scientific Assemblage Workable?
(Social Interdisciplinarity Set – Volume 2)

LARINI Michel, BARTHES Angela
Quantitative and Statistical Data in Education: From Data Collection to Data Processing
(Education Set – Volume 2)

LELEU-MERVIEL Sylvie
Informational Tracking
(Traces Set – Volume 1)

SALGUES Bruno
Society 5.0: Industry of the Future, Technologies, Methods and Tools
(Technological Prospects and Social Applications Set – Volume 1)

TRESTINI Marc
Modeling of Next Generation Digital Learning Environments: Complex Systems Theory

2017

ANICHINI Giulia, CARRARO Flavia, GESLIN Philippe,
GUILLE-ESCURET Georges
Technicity vs Scientificity: Complementarities and Rivalries
(Interdisciplinarity Between Biological Sciences and Social Sciences Set – Volume 2)

DUGUÉ Bernard
Information and the World Stage: From Philosophy to Science, the World of Forms and Communications
(Engineering, Energy and Architecture Set – Volume 1)

GESLIN Philippe
Inside Anthropotechnology: User and Culture Centered Experience
(Social Interdisciplinarity Set – Volume 1)

GORIA Stéphane
Methods and Tools for Creative Competitive Intelligence

KEMBELLEC Gérald, BROUDOUX Evelyne
Reading and Writing Knowledge in Scientific Communities: Digital Humanities and Knowledge Construction

MAESSCHALCK Marc
Reflexive Governance for Research and Innovative Knowledge
(Responsible Research and Innovation Set - Volume 6)

PARK Sejin, GUILLE-ESCURET Georges
Sociobiology vs Socioecology: Consequences of an Unraveling Debate
(Interdisciplinarity Between Biological Sciences and Social Sciences Set – Volume 1)

PELLÉ Sophie
Business, Innovation and Responsibility
(Responsible Research and Innovation Set – Volume 7)

2016

ANDRÉ Michel, SAMARAS Zissis
Energy and Environment
(Research for Innovative Transports Set – Volume 1)

BLANQUART Corinne, CLAUSEN Uwe, JACOB Bernard
Towards Innovative Freight and Logistics
(Research for Innovative Transports Set – Volume 2)

COHEN Simon, YANNIS George
Traffic Management
(Research for Innovative Transports Set – Volume 3)

EL FALLAH SEGHROUCHNI Amal, ISHIKAWA Fuyuki, HÉRAULT Laurent, TOKUDA Hideyuki
Enablers for Smart Cities

GIANNI Robert
Responsibility and Freedom: The Ethical Realm of RRI
(Responsible Research and Innovation Set – Volume 2)

GRUNWALD Armin
The Hermeneutic Side of Responsible Research and Innovation
(Responsible Research and Innovation Set – Volume 5)

LENOIR Virgil Cristian
Ethical Efficiency: Responsibility and Contingency
(Responsible Research and Innovation Set – Volume 1)

PELLÉ Sophie, REBER Bernard
From Ethical Review to Responsible Research and Innovation
(Responsible Research and Innovation Set – Volume 3)

REBER Bernard
Precautionary Principle, Pluralism and Deliberation: Science and Ethics
(Responsible Research and Innovation Set – Volume 4)

TORRENTI Jean-Michel, LA TORRE Francesca
Materials and Infrastructures 1
(Research for Innovative Transports Set – Volume 5A
Materials and Infrastructures 2
(Research for Innovative Transports Set – Volume 5B)

VENTRE Daniel
Information Warfare – 2nd edition

YANNIS George, COHEN Simon
Traffic Safety
(Research for Innovative Transports Set – Volume 4)

2015

BRONNER Gérald
Belief and Misbelief Asymmetry on the Internet

2009

VENTRE Daniel
Information Warfare

Printed and bound by CPI Group (UK) Ltd, Croydon, CR0 4YY
16/09/2026
14963089-0005